Quality Management

for the

Technology Sector

Quality Management

for the

Technology Sector

Joseph Berk and Susan Berk

Newnes
An imprint of Butterworth-Heinemann

Boston   Oxford   Auckland   New Zealand   Johannesburg   Melbourne   New Delhi

A member of the Reed Elsevier group

 Recognizing the importance of preserving what has been written, Butterworth-Heinemann prints its books on acid-free paper whenever possible.

Library of Congress Cataloging-in-Publication Data

Berk, Joseph, 1951-
  Quality management for the technology sector / Joseph Berk, Susan Berk.
    p. cm.
  Includes bibliographical references and index.
  ISBN 0-7506-7316-8 (pbk.) alk. paper
    1. Quality control. 2. Factory management. I. Berk, Susan, 1955- II. Title

  TS 156 .B467 2000
  658.5'62—dc21

                                                                    00-022363

British Library Cataloging-in-Publication Data
A catalog record for this book is available from the British Library.

The publisher offers special discounts on bulk orders of this book.
For information, please contact:
Manager of Special Sales
Butterworth-Heinemann
225 Wildwood Avenue
Woburn, MA 01801-2041
Tel: 781-904-2500
Fax: 781-904-2620

For information on all Butterworth-Heinemann publications available, contact our World Wide Web home page at: http://www.bh.com

10 9 8 7 6 5 4 3 2 1

Printed in the United States of America

This book is dedicated to the people in the factory,

and to those who support them.

Contents

Preface

TQM. MRP. JIT. ERP. SPC. DOE. ISO. TOC. CPI. CPK. CQI.

Let's face it: If you manage in a high technology environment, it may seem as though your life involves jumping from one three-letter acronym to the next.

Every management guru seems to have a new philosophy and a new set of initials he or she swears will revolutionize your company. The management fads of the last 20 years or so seem to have about a three-year half life before they start to fade away, but before their last spark, another one pops up with an accompanying new guru. There is no shortage of gurus or new acronyms, and for $1000 per day (and sometimes much more), they are happy to share their fervor with you. You spend your money and your employees' time, and a week later, you would never know you had been host to the guru-du-jour. Things look about like they did before the visit.

If you manage in the most demanding of manufacturing environments, the high technology manufacturing environment, what should you do? Should you go with TOC, TQM, or DOE? Should you get lean? Should you adopt a 5S program? Should you have a lean event? Should you opt for a Japanese-branded management philosophy for which you don't even know the English translation?

The answer is a good news/bad news story.

The good news is that many assurance technologies can make a significant improvement in the quality of the products provided by manufacturers.

The bad news is that there are no magic pills. You cannot simply buy a guru-sanctioned program (and its associated costly training and follow-on consultant support) and watch your troubles melt away. There is no substitute for informed hands-on management and leadership, and there never will be (and maybe that should be in the preceding paragraph, because we believe it is good news).

This is an unusual book. It is based on the combined observations of literally hundreds of companies making everything from biomedical devices to smart bombs, and all with one thing in common: All involved manufacturing complex products in high technology environments.

This book is different than others. It is not a touchy-feely, feel good, let's all do a better job quality management text. This book contains detailed technical reviews written in an easy-to-follow manner on basic quality management concepts, quality measurement, practical statistical techniques, experimental design, failure analysis, value improvement, supplier management, current quality standards (including ISO 9000 and D1-9000), and delivery performance improvement. The book contains many examples of high technology challenges and how people like you met those challenges. In short, this is a book for serious manufacturing managers and leaders.

Your authors have been engineering managers, quality assurance managers, manufacturing managers, and consultants to some of the largest corporations in America and overseas. This book is based on real-world observations and lessons learned by actually implementing the techniques included in the chapters that follow.

The challenges inherent to managing quality in the high technology environment are significant. No book can claim to offer a recipe for instant success in overcoming these challenges, but the approaches in the following pages can greatly ease and accelerate the quality management journey.

Chapter 1

Managing for Quality in the High Tech Environment

What American industry is doing...

Quality management in high technology environments presents a unique challenge demanding engineering, manufacturing, quality assurance, and leadership expertise. The requirements associated with high technology requirements identification and compliance, variability reduction, systems failure analysis, process control, design adequacy, cost control, and simply delivering products on time place extreme demands on managers who want to improve quality in manufacturing organizations delivering complex products. This is especially true for companies delivering cutting edge products, which typically include aerospace, defense, electronics, and biomedical manufacturers.

Let us begin our high technology quality management discussion by first understanding the concepts that guided our industrial development. These concepts are outlined in Figure 1-1.

The concept of quality control as a distinct discipline emerged in the United States in the 1920s. At the time, quality control was intended to simply control, or limit, the escape of defective items in industrial processes. As will be covered in subsequent chapters, the earliest quality control idea was to inspect the output of a manufacturing process to sort defective product from good product. There are numerous disadvantages to this sorting process, especially if the sorting is performed by different people from those manufacturing the product, but again, these concepts will be covered in far more detail later in this book.

As the quality control concept described above emerged in the first half of this century, numerous refinements occurred. Pioneering work by Shewhart, Deming, Juran, Feigenbaum, and others

indicated that there were perhaps better ways to approach quality management. Perhaps simply sorting good product from bad, they reasoned, was not the most efficient way to assure quality. A more effective management philosophy might focus on actions to prevent defective product from ever being created, rather than simply screening out defective items.

Several management theorists expanded upon this idea. Shewhart applied statistics to industrial processes in World War I. Shewhart's concept was that the use of statistical process management methods could provide an early warning, and allow the process to be adjusted prior to producing defective product. Deming and Juran based significant portions of their work on Shewhart's concept of using statistics to control processes, limit variation, and improve quality.

Quality management continued to develop under Deming's guidance, whom many regard as the father of modern quality philosophies. Interestingly, Deming's management philosophies were first developed in the years prior to World War II (not in post-war Japan, as is commonly believed). Deming believed that quality management should not focus on merely sorting good product from bad. Deming believed that the responsibility for quality should be shared by everyone in an organization. Perhaps most significantly, Deming recognized that most quality problems are system induced, and are therefore not related to workmanship.

Deming's work saw only limited application in this country prior to World War II, but a curious set of circumstances developed immediately after World War II. General Douglas MacArthur, who had been appointed military governor of post-war Japan, brought Deming to Japan to serve as a management consultant to the Japanese as they rebuilt their industrial base. Deming's message had essentially fallen on deaf ears in the United States. That did not happen in Japan.

Japan, then as now, was an island nation that had to import all of its raw materials. The Japanese were attentive listeners when Deming advised them. The Japanese saw Deming's approach as a natural approach to preventing waste. More to the point, the Japanese saw Deming's approach as a way of maximizing their productivity. Deming praised the virtues of using statistical quality control and manufacturing methods to reduce waste. Japan, as an industrialized nation that had to rebuild its industrial base from essentially nothing, absorbed Deming's teachings. The Japanese had no preconceived approaches about sorting defectives from acceptable product. They were willing to learn.

What followed in Japan during the ensuing decades has been well studied. The Japanese dominated every market they chose to enter: electronics, automobiles, steel, shipbuilding, motorcycles, machine tools, and many other products. Superior

quality became the common theme for Japanese market dominance. Much of Japan's quality superiority was based on statistical manufacturing methods. The Japanese made additional contributions to manufacturing management, most notably in the areas of variability reduction, problem solving, teams, and defining and satisfying customer expectations.

While Japan continued its quality revolution in the years following World War II, improved quality management philosophies were not pursued in the United States with nearly the same fervor as they were in Japan. The Japanese were clearly making progress in some industries, but for the most part, these inroads were not considered a serious economic threat.

The Japanese had already dominated the motorcycle industry, and they were starting to make inroads into the electronics industry. One of the largest industries in the United States, the automobile industry, was relatively untouched. The Japanese were importing a few cars to the United States, but they were much smaller than American cars and generally made no real progress into the lucrative American automobile market.

Then a significant event occurred on the world stage: The October 1973 oil embargo. Suddenly, the United States found itself wanting for oil. Small cars offering improved gasoline mileage seemed more attractive than did waiting in line for hours at the gas pump, and Americans in large numbers started seriously considering and buying Hondas, Toyotas, and Datsuns.

American consumers, to their great delight, found that Japanese cars offered significantly better gasoline mileage, but they also had another attribute: The Japanese cars were

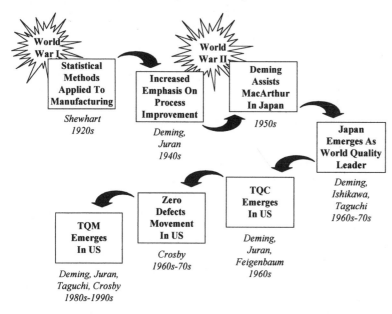

Figure 1-1. The Emergence of Quality Management Philosophies. What began as an American management philosophy died in this country, took root in Japan, and ultimately returned to flourish in the United States and other nations.

extremely well built. The quality of a Japanese automobile, especially when compared to a car produced in this country, was simply incredible.

Even though the gas crunch went away, it was too late. American drivers experienced high quality automobiles. The quality bar for automobile manufacturers had been raised, and there was no going back.

The sudden and sustained movement from American automobiles to Japanese automobiles was serious business. Up to this point, not too many people outside of Harley-Davidson really cared if you bought a Japanese product instead of an American product. When the products were cars, though, and American buyers turned to them in droves, our country began to take notice.

What has happened in the United States in the years since October 1973? Major industries (one of the first being the automotive industry) began to focus on quality in a serious manner. Other industries simply disappeared from the American landscape, succumbing to their Japanese competition (when was the last time you saw an American television, or an American watch?).

American industry is catching up, but it has been a long journey. Along the way, the United States recognized that other management philosophies should be applied to the quality improvement challenge. This blending of additional management philosophies, all targeting quality improvement, became known as the Total Quality Control concept. The concept developed under the guidance and teachings of Feigenbaum, Deming, Juran, and others. Crosby later promoted the "zero defects" concept, emphasizing adherence to requirements and employee motivation.

Total Quality Control became Total Quality Management, and that concept continued to emerge as a predominant management philosophy in the United States and abroad during the 1980s and the 1990s. TQM emphasizes a number of concepts (see Figure 1-2), all of which support the philosophies of customer focus, continuous improvement, defect prevention, and a recognition that quality responsibility belongs to each of a company's departments (not just the Quality Assurance department). Several concepts are inherent to TQM, but all support these four philosophies.

For a number of reasons, including some of those outlined in the Preface to this book, TQM's popularity has declined in the last several years. That is unfortunate, as there as several sound management philosophies and technologies that are particularly well suited to the high technology manufacturing environment. The technologies are not tied to the TQM concept, however, and in fact this book presents those and others we believe to be particularly appropriate for high technology manufacturing challenges.

Figure 1-2. The Elements of Total Quality Management. TQM is centered on the philosophies of customer focus, continuous improvement, defect prevention rather than detection, and a recognition that responsibility for quality is shared by all departments.

What are the basic elements required for managing quality in a high technology manufacturing environment? We believe they include:

- Continuous Improvement
- Customer Focus
- Quality Measurement
- Root Cause Corrective Action
- Employee Involvement and Empowerment
- Statistical Thinking
- Inventory Management
- Value Improvement
- Supplier Teaming
- On-Time Delivery Performance

Let us begin our discussion with a brief overview of these key concepts.

- *Continuous Improvement.* The continuous improvement concept simply means knowing where you are from a quality perspective and striving to do better.

- *Customer Focus*. Lee Iacocca once advertised that Chrysler had only three rules: Satisfy the customer, satisfy the customer, and satisfy the customer. That about sums up the quality management philosophy on customer focus. This philosophy is supported by a number of technologies to assure that customer needs and expectations are understood and met.

- *Quality Measurement*. Quality measurement asks the question: Where are we, and where are we going? A basic quality management concept is that quality is a measurable commodity, and in order to improve, we need to know where we are (what the current quality levels are), and we need to have some idea where we are going (or what quality levels we aspire to).

- *Root Cause Corrective Action*. Most of us have experienced instances in which problems we thought were corrected continued to occur. The problem is particularly vexing in the high technology environment. Problems in complex products are difficult to define and to correct. There are several technologies associated with this endeavor. One consists of basic problem solving skills, another consists of a more advanced systems failure analysis approach, and still others involve statistical analysis and designed experiments.

- *Employee Involvement and Empowerment*. Employees must be involved and empowered in high technology manufacturing environments. Employee involvement means that every employee is involved in running the business and plays an active role in helping the organization meet its goals. Employee empowerment means that employees and management recognize that many obstacles to achieving organizational goals can be overcome by employees if they are provided with the necessary tools and authority to do so.

- *Thinking Statistically*. Statistical thinking is a basic requirement when managing quality in a high technology environment. Quality

improvement often requires reducing process or product design variability reduction, and statistical methods are ideally suited to support this objective.

- *Inventory Reduction*. Largely in response to their lack of natural resources (as well as the 1970s worldwide oil shortages), the Japanese pioneered the concept of reducing inventories. This management philosophy became known as Just-In-Time (or JIT, for short) inventory management. Although the concept was originally intended to address material shortages, an interesting side effect immediately emerged: As inventories grew smaller, quality improved.

- *Value Improvement*. There is a linkage between continuous improvement and value improvement that is simultaneously obvious and subtle. This linkage becomes apparent when one considers the definition of quality, which is the ability to meet or exceed customer requirements and expectations. The essence of value improvement is the ability to meet or exceed customer expectations while removing unnecessary cost. Removing unnecessary costs while simultaneously satisfying customer expectations and requirements can only serve to increase customer satisfaction (after all, the customer is receiving the same level of quality for a lower cost).

- *Supplier Teaming*. Another philosophy inherent to managing quality in a high technology environment is that of developing long term relationships with a few high quality suppliers, rather than simply selecting those suppliers with the lowest initial cost. American industry and government procurement agencies have had, and are continuing to have, difficulty in implementing this concept, although progress is being realized.

- *On-Time Delivery Performance*. One of the most common complaints manufacturing organizations (and their customers) have about their suppliers is that they cannot deliver products on schedule. If we accept the notion that quality is defined by meeting customer requirements and expectations, then we have to realize that delivering on time is a key customer satisfaction index. We devote an entire chapter to this subject at the end of this book. On-time deliveries are key to earning and keeping satisfied customers.

Summary

Quality management in the high technology manufacturing environment presents unique challenges. Quality management is not a discipline that can be delegated to an organization's Quality Assurance department; rather, it is responsibility that is shared by all. This is particularly true for manufacturing managers.

Many of our quality management disciplines go back nearly a century, with others emerging more recently. These technologies developed largely as the result of pioneering work by Deming, Juran, Shewhart, Feigenbaum, and others. More sophisticated manufacturing and quality management concepts (primarily those based on statistical thinking and focusing on the customer) did not immediately take root in the United States, but they did in Japan in the years following World War II. As a result Japan emerged as a world quality leader. The United States has made significant inroads and in many regards has surpassed Japan in high technology quality in manufacturing organizations. The technologies supporting manufacturing management in high technology organizations emphasize a number of management concepts, all of which are centered on philosophies of customer focus, continuous improvement, defect prevention, and a recognition that responsibility for quality is shared by all.

References

"Small Firms Put Quality First," *Nation's Business*, Michael Barrier, May 1992.

"The Cost of Quality," *Newsweek*, September 7, 1992.

"Six Sigma: Realistic Goal or PR Ploy," *Machine Design*, Jim Smith and Mark Oliver, September 10, 1992.

Commonsense Manufacturing Management, John S. Rydz, Harper and Row, Inc., 1990.

Thriving on Chaos, Tom Peters, Alfred A. Knopf, Inc., 1987.

The Deming Management Method, Mary Walton, Perigee Books, 1986.

Quality Is Free, Philip B. Crosby, McGraw-Hill Book Company, 1979.

Guide to Quality Control, Kaoru Ishikawa, Asian Productivity Organization, 1982.

Thriving on Chaos, Tom Peters, Alfred A. Knopf, Inc., 1987.

The Deming Management Method, Mary Walton, Perigee Books, 1986.

Chapter 2

The Continuous Improvement Concept

Initiating a continuing journey...

Steve Michaels studied the Pareto charts in front of him. Michaels was an assembly area supervisor in Parsons-Elliason, a company that developed and manufactured mass spectrometers. He had been challenged by his boss, Ed McDermitt, to find the top three areas in the company requiring improvement. Michaels found the top three items on the Pareto chart that listed nonconformances by quantity, and the top three areas on the chart that listed nonconformances by cost. The two charts did not match. The top three high count nonconformances did not match the most costly nonconformances. Michaels felt ready to take his suggestions to the boss. He would recommend attacking the high cost items first.

"Good afternoon, Steve," McDermitt said when Michaels entered the office. "What's up?"

"I've got some suggestions on the question you asked me yesterday," Michaels said.

"And that question was?" McDermitt asked.

"You wanted what you called continuous improvement suggestions," Michaels said. "I've looked at the Pareto charts the quality guys prepared, and I picked the three most expensive components, in terms of what these failures are costing the company."

"Okay, that's good," McDermitt answered. "What are your suggestions?"

"There's a power supply we buy from Paradyne Products that's the most expensive one," Michaels said. "I recommend we find a new supplier, because Paradyne's power supply units are failing frequently enough to be number one on the cost chart. The

other two are circuit card assemblies we buy from Lampson Electronics. They aren't failing as often, but they're expensive, too, and I recommend we go to a new vendor."

"How do you know these parts are bad?" McDermitt asked.

"They show up as the most expensive failures," Michaels answered. "That's what this continuous improvement business is all about, right? Find the biggest problems, fix them, move on to the next biggest problems, fix them, and so forth."

"You've got the right idea," McDermitt answered, "but you may want to consider other options before we drop these guys. We worked with Paradyne and Lampson a long time to get power supplies and boards that meet our requirements. Maybe the problem isn't with their equipment, but it's got something to do with how we handle them once they get here instead. Have you talked to the people who install these things to see what they think?"

"Well, no," Michaels answered. "You think we could be causing the problems?"

"I don't know," McDermitt said. "Talk to the guys in the shop. You've got the right idea. We need to fix whatever it is that's causing the power supply and Lampson board failures, but it may not be the supplier's fault. There might be something in our assembly process that's causing the problem. But keep at it, and you're right about continuous improvement. When you fix these problems, we'll move on to the next ones."

Michaels got up to leave, but McDermitt spoke again.

"You know, there really is more to the continuous improvement concept," McDermitt said. "What do you think our objectives ought to be?"

"What do you mean?" Michaels said. "I think if we fix these three problems, we'll just move on to the next ones."

"Yes, I agree with you on that," McDermitt said. "But how do we know if we're really getting better? I mean, suppose that we continue to have other problems just as severe, or just as expensive. Problems that pop up when these go away. Would we really be getting any better?"

"I don't understand," Michaels said.

"Go upstream," McDermitt said. "In addition to fixing the problems with the power supplies and the circuit cards, why don't you take a look at how we came to have these problems? Perhaps we aren't doing something right in the way we design or specify components. Perhaps we don't inspect them adequately, and the way we define the inspection requirements isn't good enough. Take a look at the whole process. We don't want to fix these problems just to have three others pop up that are just like them. Oh, and get some help. See if you can put together a team that might have other insights into the big picture."

A Quality Management Foundation

Continuous improvement is an inherent part of the quality management process. Continuous improvement consists of measuring key quality and other process indices in all areas, and taking actions to improve them. These indices could include the output of a manufacturing process, customer satisfaction, the number of engineering drawing errors per month, warranty returns, or any of a number of other measures used to characterize a process. As the definition states, continuous improvement should be focused on processes, and pursued in all areas. The continuous improvement concept focuses on finding shortfalls and sources of variability in administrative, manufacturing, and service processes that can detract from a quality output, and improving the process to eliminate undesirable outputs.

What is a process? A process is a series of activities by people or machines that move work toward a finished product. The objective of continuous improvement is to improve the process such that customer satisfaction increases, and the cost of attaining this increased customer satisfaction decreases.

The Continuous Improvement Approach

How does one go about implementing continuous improvement? Figure 2-1 shows a strategy we prefer and have used successfully in a number of organizations.

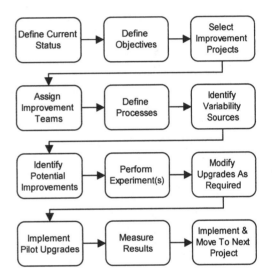

Figure 2-1. A Strategy for Implementing Continuous Improvement. The path outlined above provides a good road map for realizing continuous improvement.

The continuous improvement process begins by defining an organization's current quality status. We'll see how one goes about doing this in Chapter 4, which discusses quality measurement systems. The concept in this first continuous improvement step is to identify an organization's current quality status. This can be addressed from any of several perspectives, including number of defects, the cost of defects, customer satisfaction indices, and perhaps other indices. The measurement indices used to determine an organization's quality status are unique to the type of business, and frequently, to the organization itself.

Defining Continuous Improvement Objectives

Once the organization's current quality status is known, the next step is to select continuous improvement objectives. The first step asked the question: Where are we? This second step asks the question: Where are we going? When pursuing continuous improvement, an organization's quality improvement objectives should be based on a realistic appraisal of what the organization, with its available resources, is capable of attaining. Establishing unrealistically high continuous improvement objectives invites failure, and that can have a demotivating effect. Our experience indicates it's better to set modest improvement goals at first so that a few successes can be realized. These initial successes will help others in the organization buy into the continuous improvement philosophy.

Converting Objectives into Actions

The next step is to convert the continuous improvement objectives into action, and that means selecting continuous improvement projects. These are the specific areas in which an organization desires to seek improvement. Perhaps a product fails too often during acceptance testing, and the goal is to reduce test failures by 50 percent. Perhaps the finance department is habitually late in paying accounts payable, and the goal is to assure all payments are made in less than 30 days. Perhaps work instructions contain too many errors, and the goal is to cut work instruction errors to less than one-tenth of current values. Each of these projects provides the framework of an action plan for the organization to realize continuous improvement.

People Make It Happen

Having selected areas in which to focus continuous improvement efforts, the organization next has to assign people to work these projects, and empower them to attain continuous improvement objectives. Chapter 7 provides strategies for employee involvement and empowerment. Chapter 8 presents a framework for tailoring teams based on the nature of the continuous improvement project and other parameters. These concepts of involvement, empowerment, and teams are extremely important to realizing continuous improvement, as they allow an organization to attain significant synergies and fully utilize its human resources.

Process Definition

Once the team has been assigned to a continuous improvement project, it should begin by defining the process it is assigned to improve. We recommend preparing simple flow charts for this purpose (an example is included in Figure 2-2). This concept of flow charting processes will be further developed in

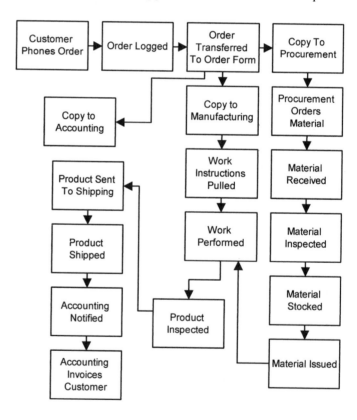

Figure 2-2. Order Processing Flow Chart. Flow charting is a good way to define a process, to gain insights into problem areas and inefficiencies, and to develop continuous improvements. The flow chart shown here, if carefully studied, can reveal unnecessary actions and several sources of variability.

Chapter 10 (on statistical process control), and in several other chapters as well.

Preparing flow charts to define processes (whether they are for creating engineering drawings, manufacturing a product, administering a performance appraisal, or any other process) is often an eye-opening experience for the people involved. We've observed many surprised people (including those who managed and worked as part of the process being flow charted) during this exercise. Many people who manage or work in a process don't realize what makes up the entire process. Flow charting provides this visibility. Flow charting also often shows many problem areas and inefficiencies. People who work in the process often can't see the forest for the trees, and putting the process on paper helps to eliminate these blinders.

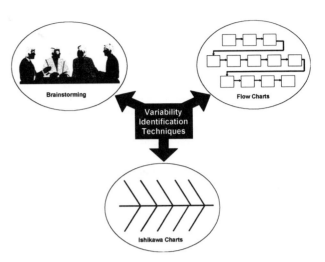

Figure 2-3. Popular Variability Identification Approaches. Brainstorming, Ishikawa cause-effect diagrams, and flow charts all serve to identify sources of variability. Variability reduction results in improved quality.

What does one look for in process flow charts? For starters, human inputs should be identified. Wherever a human input is required, potential sources of variability can enter the process. To achieve continuous improvements in a process, one should take steps to clarify or limit the human inputs to control this source of variability. Blocks that go nowhere (for example, the "copy to accounting" block in Figure 2-2) generally reveal unnecessary actions. Finally, each step should be examined, and the team should ask the question: What happens if this step is eliminated? If the answer is nothing, the step should be removed from the process.

Variability Reduction Equals Quality Growth

Having defined the process under study with the aid of a flow chart, the continuous improvement process next moves on to defining areas in which variability can creep into the process. Another TQM concept is variability reduction, and the thought that anything done to reduce variability results in improved quality. Problem solving, systems failure analysis, statistical process control, Taguchi philosophies, and supplier teaming all serve to reduce variability, and the chapters on these subjects develop technologies for variability reduction. Three of the most common variability identification approaches are simple

brainstorming among the team members, Ishikawa cause-effect diagrams (these will be covered in Chapter 10), and taking a hard look at the process flow chart to identify where variability can enter the process (see Figure 2-3).

As Chapters 9 and 10 will explain, there are two sources of variability present in every process. One is normal variability, which is due to the randomness associated with the process. The other is special variability, which is induced by something not controlled in the process. Variability reduction aims to make sure the normal variability inherent to a process is not so great that the process will produce a product that exceeds its specification limits, and that the causes of special variability are eliminated.

This concept of process improvement through variability reduction is key to successful quality management implementation. Deming taught that fully 85 percent of an organization's quality deficiencies are due to the variability induced by process problems, and not workmanship. To gain the most from a continuous improvement effort, it makes sense to focus on process improvement. Instead of finding someone to blame when things go wrong (or limiting the application of a corrective action to fix a specific defect), we believe good manufacturing and quality managers instead zero in on the process deficiencies that allowed the problem to develop. The idea is that eliminating process deficiencies and minimizing process variability will prevent future defects.

Once the sources of variability have been identified, potential improvements can then be developed. Again, teams offer more than do individuals working in isolation. Chapters 5 and 6 (on problem solving and failure analysis) offer approaches for developing potential corrective action solutions for continuous improvement and problem prevention.

Implementing Change: Managing the Risk

Good risk management mandates a thorough evaluation of any process improvements prior to implementation, and the next four steps in the continuous improvement process serve to mitigate the risk associated with any process modification. We recommend designing tests or experiments (when practical to do so) to evaluate the feasibility of any process modification. These tests will show if the process modification will work, and any required modifications prior to implementation. We also recommend that whenever possible, the process upgrade be incorporated as a pilot program in a small area prior to full implementation. For example, if you work in a manufacturing environment and a continuous improvement team recommends modifying the way your organization issues material to the shop floor, it would make sense to try this in a small area of the plant prior to full factory implementation. The pilot program will identify risks associated with proposed process modifications, and where problems emerge, they can be corrected prior to full implementation.

We recommend monitoring the pilot program process upgrades using the same measurement criteria that initially targeted the process for improvement. This will help to determine if the

process improvement actually resulted in an improvement. We also recommend continuing to monitor the process with the same measurement criteria once the upgrade has been fully implemented.

What happens after the process improvements have been implemented and confirmed as effective? As the name implies, an organization implementing continuous improvement moves on to the next project to realize additional continuous improvement gains. The process never ends.

Summary

Continuous improvement consists of measuring key quality and other process indices in all areas, and taking actions to improve them. Continuous improvement should be focused on processes and pursued in all areas. The continuous improvement concept focuses on finding shortfalls and sources of variability in administrative, manufacturing, and service processes that can detract from a quality output, and improving the process to eliminate undesirable outputs.

References

A Guide for Implementing Total Quality Management, Reliability Analysis Center, United States Department of Defense, 1990.

Delivering Quality Service, Valarie A. Zeithaml, A. Parasuraman, and Leonard L. Berry, The Free Press, 1990.

Applications of Quality Control in the Service Industries, A.C. Rosander, Marcel Dekker, Inc., 1985.

A Handbook for First-Time Managers: Managing Effectively, Joseph and Susan Berk, Sterling Publishing Company, 1997.

Chapter 3

Finding Your Customers

Everyone serves internal and external customers...

Tom Axelson shook his head in disbelief. As Defense Systems Associates' program manager for the AN/RPV-39 air vehicle, he stared at the telefaxed letter in front of him. The fax paper held every defense contractor's nightmare: a "show cause" letter. The message from the U.S. Army contained a single and painfully blunt sentence:

"Based on Defense Systems Associates' inability to deliver AN/RPV-39 Remotely Piloted Reconnaissance Vehicles on schedule and in a condition that meets performance specification requirements, this office directs that Defense Systems Associates, within the next 10 days, show cause as to why this contract should not be terminated for default."

Axelson had recognized the situation was serious for the last several months, but the message in front of him was sobering. The United States Army was telling Defense Systems Associates that unless it could show adequate reasons for the company's poor performance, the Army would cancel a contract worth in excess of $60 million.

Axelson thought back to the euphoria that had swept over Defense Systems Associates when they first won the AN/RPV-39 development and production contract two years ago. As a small technology-oriented company, Defense Systems Associates had experienced annual sales of approximately $12 million for several years. Winning a competitive, multi-year program virtually assured Defense Systems Associates' survival in a shrinking industry. The Army's new remotely piloted tactical reconnaissance program had been one of the few defense industry windfalls from the Persian Gulf war, which demonstrated gaps in the military's

capability to secure rapid information on enemy troop movements and other activities.

Defense Systems Associates had built remotely piloted reconnaissance vehicles for the Army and the Marine Corps in the past, but the earlier contracts had been for relatively unsophisticated single vehicles involving low technology camera systems (none of the prior contracts had exceeded a million dollars). The AN/RPV-39 was a much more complex vehicle, with television and infrared cameras, electronic eavesdropping equipment, and data links to provide information on the enemy as soon as the air vehicle detected it. The AN/RPV-39 contract offered Defense Systems Associates financial growth and a chance to significantly enhance its technical staff and manufacturing capabilities. The program moved the company into a dominant position in an industry that previously held no clear leaders.

As one of the company's brightest engineers, Tom Axelson had been selected to manage the AN/RPV-39 program for Defense Systems Associates. As the AN/RPV-39 program manager, Axelson was responsible for building a team of engineers, manufacturing engineers, quality assurance experts, procurement specialists, and other engineering and manufacturing professionals. Axelson's charter was to lead his team to first design the system, build two prototypes (which would ultimately be delivered as production vehicles), and then build three more of the remotely piloted reconnaissance aircraft. The first two prototypes were to be designed, built, tested, and delivered to the customer 18 months after the contract had been signed.

Axelson looked at his calendar. The AN/RPV-39 contract had been signed 25 months ago, and the company had yet to complete successful testing on the two prototypes that remained parked in the Defense Systems Associates hangar. The problems emerging during the development phase of the AN/RPV-39 program seemed endless, as did the arguments and ill will. Hostile feelings between Defense Systems Associates and the Army were rampant, as were similar feelings between individuals and departments within Defense Systems Associates. Axelson had never worked on a program that seemed to generate so many personality conflicts.

Axelson took the letter to his boss, Aldo Pietras, the president of Defense Systems Associates. Pietras smiled when Axelson walked into his office, but when Axelson placed the letter in front of Pietras and he read it, he, too, was stunned. Pietras had formed Defense Systems Associates 22 years before, and had nurtured the company's development through the post-Vietnam defense industry cutbacks.

Pietras read the brief letter twice before commenting. "Those kids in the Army think they know how to run a program. They send us a letter like this...they ought to be ashamed of themselves. They're the ones that are causing these problems, with their ridiculous performance specifications. There hasn't been a week gone by that they haven't changed the requirements on us."

Axelson stared at the floor. He knew that Defense Systems Associates wrote the performance specifications for the Army before they won the contract. The Army wanted the AN/RPV-39 aircraft to do everything in the performance specifications, but only because Defense Systems Associates had assured the Army the aircraft could meet the requirements. Axelson also knew that the Army's specification changes had been rational, and the company had agreed with them.

"We're having more problems inside the company, too," Axelson said to Pietras. "Everyone is upset with everyone else. We're practically having a war between Engineering and Manufacturing. Manufacturing claims the design is too difficult to manufacture, and Engineering thinks the people in Manufacturing are incompetent. Our people can't work with each other even within Manufacturing.

The sheet metal assemblers are complaining that the panels they receive from the stamping area are not built to print, and they have to be reworked before they can be used. The stamping people don't seem to care. The composites layup people have given up on both groups." Axelson looked up at Pietras.

"They're all wrong," Pietras answered. "Our stampings are the best in the industry. So are our design and our engineering people. And no one has a better group of assembly people."

"That may be, sir," Axelson said, "but you couldn't see it if you came in from the outside, which is how the Army is seeing us. It's almost as if no one in the plant cares about the next guy down the line." Axelson paused, concerned that he might have overstepped his bounds with Pietras.

"What do you mean?" Pietras asked.

"Well," Axelson began, "inside the company, no one seems to give a damn about the person, or group, that will be using whatever it is they make." He paused, looking at Pietras. Pietras had a reputation for shooting the messenger. Axelson was afraid that Pietras was offended by his comments.

"Go on," Pietras said.

"Engineering creates a design that Manufacturing has to build," Axelson continued. "Manufacturing says they can't build to the engineers' design, but the engineers don't listen. The sheet metal assemblers complain about the quality of the sheet metal stampings, but the stamping supervisor doesn't do anything to improve the quality of her group's output. She doesn't even seem to recognize that there is a problem. And as a company, we don't seem to get too concerned about what our final customer, the Army, wants. The Army gets upset because we are more than six months behind schedule and they send us a show cause letter, and our first reaction is that the Army is wrong. It just seems that we are not paying attention to what the customer wants, both our internal customers, and externally, with the Army."

Pietras sat up and looked at Axelson. "You know," he said, "you just might be on to something. Please continue."

Pietras stared through his window for a moment before continuing. "I've sensed the same thing myself, although I couldn't articulate my thoughts as clearly as you just did. We have forgotten that we are here to serve the customer, whoever that is. We have lost sight of what a customer is, both internally and externally. That is our problem. I used to think that all of the difficulties we have experienced recently, what we have been going through, was a natural fallout of a company's growth, but now I don't think so. Our company has forgotten why we are here. We're here to build the best reconnaissance vehicles in the world, meet our customers' needs, and make a profit in the process. I've probably contributed to this failure myself by not demanding that everyone recognize that our jobs depend on satisfying the customer, and by being too quick to blame others for our shortfalls."

Pietras stopped and looked at Axelson. Axelson was stunned. He had never heard Pietras be so self-critical and honest about the company's situation. The show cause letter was obviously a significant emotional event.

"What do you recommend we do?" Pietras asked.

What Is a Customer?

Webster defines a customer as "one that purchases a commodity or service." That definition provides a start, but it needs to be developed from a quality management perspective. Webster's definition implies an interface between two individuals or organizations, in the sense that one sells to the other (Figure 3-1 shows the concept). That fits the definition of but one type of customer. For the purposes of this discussion, the concept of two categories of customers is helpful: the external customer and the internal customer.

External Customers

External customers are what Webster probably had in mind in formulating his definition. These are the people or organizations that buy what an individual or an organization sells. The concept is simple enough to be illustrated by a few examples:

- A person buys a car from a new car dealer (that person is the new car dealer's customer).

- A couple have dinner at an exclusive restaurant (the couple are the restaurant's customers).

- A consultant prepares a market trend analysis for a motorcycle manufacturer (the motorcycle manufacturer is the consultant's customer).

- A defense contractor manufactures a weapon system for the Department of Defense (the Department of Defense is the defense contractor's customer).

- Defense Systems Associates is under contract to develop and manufacture reconnaissance vehicles for the United States Army (the Army is Defense Systems Associates' customer).

In the context of Webster's definition, external customers are those outside the bounds of an organization who buy what the organization sells.

Figure 3-1. Traditional Customer/Supplier Concept. In this concept, as defined by Webster, the customer and the supplier are distinct entities, with the supplier selling goods or services to the customer.

Internal Customers

Here's where the concept of the customer becomes a little more complicated, but only slightly so. Let's take Webster's definition of a customer (which is how most of us think of customers) and modify it slightly. Instead of defining a customer as one who purchases a commodity or service, let's instead call a customer anyone (or any organization) that receives and uses what an individual or an organization provides. This definition has significant implications. Based on it, customers are no longer necessarily outside the bounds of an organization selling a commodity or service. To be sure, every one of the customers cited as examples above still fits our modified definition, but note that an entirely new category of customers can emerge.

These customers are significantly different than the customers presented as examples in the preceding pages. Instead of being outside of the organization supplying the goods or services, these customers can be inside the organization doing the supplying (i.e., the selling organization). Figure 3-2 shows the concept.

Consider with us a relatively simple example in a manufacturing environment. Let us examine an assembly line producing recreation vehicles, and in particular, those portions of the assembly line that mount tires on wheels and install the wheels on the RV coaches.

As Figure 3-3 shows, the assembly line has numerous other work groups performing various specialized tasks, but if

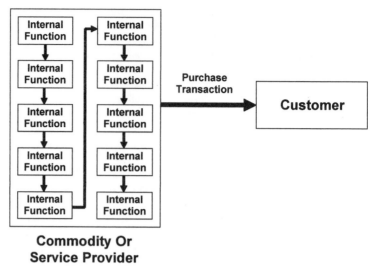

Commodity Or Service Provider

Figure 3-2. Internal and External Customers. Note that the commodity or service provider provides its product to the external customer, but there are numerous internal functions within the supplier. Each function is an internal customer of those functions that precede it in the process of preparing the goods or services to be provided to the external customer.

the focus is on just the two work groups described above, it becomes clear that the work group mounting the tire on the wheels is providing a product to the work group that installs the wheels (with tires) on the coach. The first work group can be thought of as a supplier, an organization devoted to meeting the needs of its customer. The first work group's products are complete wheel and tire assemblies (i.e., wheels with tires properly mounted). The wheel installation work group can be thought of as a customer. The wheel installation work group receives the product of the wheel and

tire assembly work group, thereby meeting the requirements of our modified definition of a customer.

What are the implications of this new customer definition, and the concept of an internal customer? From the perspective of a manufacturing organization, the implications are far-reaching. Consider the following questions:

- Do the wheel and tire assemblies provided by the wheel and tire assembly work group have to meet the needs and expectations of the wheel installation work group?

- What happens if the wheel and tire assembly work group does not satisfy the wheel installation work group?

- Do the wheel installation work group's requirements fall within the capabilities of the wheel and tire assembly work group?

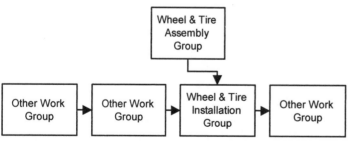

Figure 3-3. RV Assembly Line Internal Customers. The wheel and tire installation group is an internal customer of the wheel and tire assembly group.

- If the wheel and tire assembly work group provides a low quality product (improperly mounted tires, unbalanced wheel assemblies, or otherwise damaged wheel assemblies), what does that do to the output of the wheel installation work group? What does it do the recreational vehicle manufacturer's external customers (those who ultimately purchase the automobiles)?

- Is the cost of doing business with the wheel and tire assembly work group higher (either in monetary terms, or in terms of lower quality, and therefore, more rejected wheel assemblies) than the cost of buying completed wheel assemblies from an external supplier?

Suppose the wheel and tire assembly work group provides wheel assemblies of low quality that do not meet the needs and expectations of the wheel and tire installation work group. Will the wheel and tire installation group be dissatisfied with the goods it is receiving? Will it be unable to use these wheel and tire assemblies? Will the output of the wheel and tire installation work group suffer if it receives low quality wheel assemblies?

The answer to each of the above questions is yes, and that has strong implications for the last question. As is the case any time a customer is dissatisfied (in this case, the wheel and tire installation group), the supplier is likely to lose business to a competitor. Although it may sound incredible, many companies in the United States are outsourcing work previously done internally for just that reason. They have discovered that external suppliers can often provide higher quality goods and services than can be provided internally.

Selling Is Not Always Required

As the above example shows, a group serving the needs of its customers may not necessarily sell its products. The customer can be an internal customer that does not engage in trade, but instead simply receives the output of its suppliers (this occurs most often in relationships involving internal suppliers and customers, as is the case in the example provided above).

The lack of selling as a sign of a customer/supplier relationship is not confined solely to internal supplier/customer relationships, however. Consider the goods and services provided by government agencies and social services. The services of local police and fire departments are provided to customers (in this case, the inhabitants and visitors of the areas served by the police and fire department). If poor service is provided (perhaps the agency takes too long to respond to calls for assistance), the agency's customers will be every bit as dissatisfied as a customer who bought a product or a service.

There are numerous other government agencies and social services that serve customers. Government bodies and social agencies recognize that their services must meet the needs and expectations of their customers. Many police departments are now tracking their indices of customer satisfaction, such as emergency call response times, crime statistics, and the time it takes to solve crimes (and they are taking actions to improve these statistics). A few police departments are even mailing cards to crime victims to advise them on the status of the investigation, who the investigating officers are, and how they can be contacted for additional information. Some motor vehicle registration services are now tracking the average amount of time drivers must stand in line to register vehicles or wait for state-mandated inspection services (and they are taking actions to reduce these delays). Some community health services query those to whom they provide services (and they are moving to improve areas where their customers are not satisfied). Many schools are now tracking average test scores of their students on standardized tests (and taking actions to improve student skills in weak areas).

Everyone Has a Customer

Based on all of the above, it becomes apparent that virtually anyone engaged in any organized endeavor has a customer. Assembly line workers provide goods or services for the next worker on the assembly line. Workers sanding metal surfaces prior to painting are providing a service for those in the paint shop. Typists provide word processing services to whose for whom they type. Maintenance and janitorial personnel provide services that keep buildings clean and in good operating order for those who work in the building. Stock clerks keep goods neatly on display for those who sell to the public. Teachers provide learning experiences for their students. Automobile dealers sell and maintain automobiles for their customers. Defense contractors sell to Defense Department procuring agencies. Defense Department procuring agencies manage the development and procurement of weapons systems for military users. Military organizations maintain a high state of combat readiness to deter others from infringing upon our national interests. Everyone has a customer.

Identifying and Satisfying Your Customers

To best satisfy the needs of a customer, it almost goes without saying that one providing goods and services must know to whom the goods and services are being provided, and what their needs and expectations are (this goes back to our earlier definition of quality). We believe that most cases involving dissatisfied customers occur when those providing goods or services fail to understand the needs and expectations of their customers, or they fail to understand who the customer is. Chapter 12 will present a sophisticated methodology for thoroughly developing customer needs and expectations (the approach is called Quality Function Deployment). For now, let's concentrate on simply identifying who the customer is and what the customer wants.

Who Is the Customer?

This may seem as if it is a fairly simple question, but it often is not. Consider external customers first. In many instances, one might consider that external customers are obvious. A person who buys a new car is clearly the customer. A man buying groceries is clearly the customer. Perhaps, however, the answer is not so obvious. Is the person buying the new car the only customer? What about the person's family, or others who might travel in the car? Consider the man buying groceries again. Is he the customer, or is the customer the people he will cook for, or are both customers?

The situation is further complicated when one provides goods or services to an organization. Larger organizations typically have purchasing groups that buy for the entire organization, or for another department within the organization.

Suppose Defense Systems Associates, the organization described at the beginning of this chapter, is considered again. The group within the Army that wrote the show cause letter is most definitely a customer, but they are not the only customer. They are buying for the soldiers who will ultimately use the Defense Systems Associates' aerial reconnaissance systems. To better serve the needs of their customers, Defense Systems Associates will need to understand not only the needs and expectations of the procuring agency (i.e., the group buying the product), but also the needs and expectations of the soldiers who will use the system. Our experience has proven that when selling to organizations, it is not unusual to have many more customer organizations within the buying organization.

Given the above situation (which typically exists when serving large organizations), evaluating the customer organization makes good business sense. This analysis involves identifying lower-tier organizations and individuals within each lower-tier organization (all within the larger customer organization). Once this exercise is complete, one can identify the needs and expectations of each individual within the customer community. Many organizations serving large customer organizations go one step further and identify the individuals (within the serving organization) who need to maintain an interface with their customer counterparts.

Figure 3-4 on the next page shows an example of a customer counterpart identification matrix that Defense Systems Associates might have prepared. A matrix of this type can be used for outlining who the customers are in a large organization like the United States Army, and who their counterparts are in an organization like Defense Systems Associates.

Identifying Internal Customers

If the above seems complicated, consider how difficult it can be to identify internal customers. How is one supposed to know who the internal customers are?

Fortunately, the situation is not as complicated as it

Defense Systems Associates		US Army		
Name	**Title**	**Name**	**Title**	**Organization**
Aldo Pietras	President	BG R. Hollenbeck	Aviation Systems Command Commander	Aviation Systems Command
Tom Axelson	Program Manager	LTC F. Carpenter	Program Manager	Aviation Systems Command
		LTC W. Jordon	Squadron Commander	United States Eighth Army
Tom Rivera	Project Engineer	MAJ E. Ferlingen	Project Engineer	Aviation Systems Command
Orlando Adams	Quality Manager	LTC R. Leskewiecz	Quality Assurance Director	Aviation Systems Command
Lynn Bedelin	Contracts Administration	Ms. G. Handleson	Procuring Contracting Officer	Aviation Systems Command
Tom Granger	Subcontracts	Ms. G. Handleson	Procuring Contracting Officer	Aviation Systems Command
Bob Nakasa	Test Engineer	CPT D. Donaldson	Test Coordinator	Aviation Systems Command
Bill Boyd	Manufacturing	Mr. A. Foley	Producibility Manager	Aviation Systems Command
Kim Patterson	Manufacturing Engineering	Mr. T. Dellinger	Producibility Engineer	Aviation Systems Command

Figure 3-4. A Customer Counterpart Matrix. This chart was prepared for Defense Systems Associates to show individual customers within the customer organization (in this case, the U.S. Army), and their supplier counterparts in Defense Systems Associates.

might seem. Identifying customers can be as simple as referring to the definition developed earlier, and asking the question: Who receives or is influenced by the product or service my group provides? Simply identifying those who are affected by a product or service will reveal who can be pleased or displeased by it. Those people are your customers.

Satisfying the Customer

Satisfying the customer is a simple concept. It involves defining the customer's needs and expectations, and then meeting those needs and expectations. As mentioned earlier, fully developing those needs can be as complex as performing a quality function deployment analysis, or as simple as listening to the customer. Satisfying the customer has to be a paramount concern, as dissatisfied customers represent major lost opportunities. Recent studies confirm that developing new customers to replace dissatisfied customers costs an average of five times more than it does to retain a satisfied customer. Other studies

show that dissatisfied customers express their dissatisfaction to an average of 11 other potential customers. From any perspective, the cost of a dissatisfied customer extends well beyond the business lost to that single customer.

Dissatisfied Customers: Hidden Opportunities

All businesses, no matter how hard they work to please their customers, face the problem of dissatisfied customers from time to time. That should not be a cause for concern. Numerous studies show that when a customer is dissatisfied and the situation inducing the dissatisfaction is reversed, the customer will remain loyal and spread the news about the extraordinary actions an organization or individual undertook to eliminate or otherwise rectify the source of dissatisfaction.

An example might help to further develop this point. A colleague of ours bought a new automobile, and during its first warranty service the dealer dented one of the doors. The dealer apologized and repainted the door, but the paint did not match the rest of car (at least to our friend's satisfaction). Our friend complained, convinced that his car and its repaired door would forever be a source of irritation. Much to our friend's amazement, the dealer offered a new car in exchange for the one with the repainted door. Did our friend's dissatisfaction disappear? Absolutely. Did he buy another car from the same dealer? As a matter of fact, he bought three over the next ten years. Did our friend tell others about the dealer's commitment to customer satisfaction? The answer is an emphatic yes. Is it likely the dealer gained new business from others who heard the story? You be the judge. Our guess is that he did.

The above situation need not be confined to individual consumers. Aerojet Ordnance won an $80 million Tri-Service build-to-print production contract for the Gator weapon system, only to fail the first article flight test. Aerojet's failure analysis showed conclusively that the government's design was seriously flawed. Aerojet could have paused and waited for the government to compensate the company for its expenses in the failed flight test. Instead, Aerojet pressed ahead and corrected the government's design deficiencies, passed the flight test, and delivered the highest reliability mine systems the services had ever procured. Chuck Sebastian, the Aerojet president, made a wise business decision. Over the next 10 years, the government awarded Aerojet hundreds of millions of dollars in new Gator business.

Unearthing Sources of Dissatisfaction

Neither organizations nor individuals can afford to be passive when identifying sources of customer dissatisfaction. A strong proactive stance is required. No one can assume that just because there are no complaints the customer is satisfied. Research in this area shows that most dissatisfied customers will not complain to the organization or individual with whom they are dissatisfied. The research also shows that nearly all dissatisfied customers will never return with additional business. They will, however, tell other potential customers about their dissatisfaction.

How does one go about measuring customer satisfaction? Some companies provide questionnaires to find out how well they are doing.

Others send surveys through the mail to the customers. While these actions are often good indicators of customer satisfaction, they are somewhat impersonal. Some believe if a customer is dissatisfied, asking the customer to fill out a form to register his or her dissatisfaction tends to exacerbate ill feelings. Most agree that the human touch is far more revealing. Simply asking "How are we doing?" will often elicit a meaningful response. In our experience, this happens far too infrequently. (Have you ever noticed that restaurants with excellent service always ask customers if they enjoyed their dinner, while those with poor service never ask?) One final suggestion is a follow-up phone call, either to ask about the customer's satisfaction or to follow up on a complaint. This almost always elicits a favorable reaction, and sometimes even results in additional business.

Revisiting Defense Systems Associates

Let's now return to the organization described at the beginning of this chapter, Defense Systems Associates (an organization with obviously dissatisfied internal and external customers). What can Defense Systems Associates do to turn a bad situation around?

The management of Defense Systems Associates understands that they have a serious situation on their hands. Recognizing that Defense Systems Associates is selling to a large organization (the U.S. Army), one of their first tasks must be to recognize who within the Army is dissatisfied. At this point, it might include everyone in the Army who is associated with the AN/RPV-39 program. Nonetheless, Defense Systems Associates needs to identify who their Army customers are (in accordance with the customer identification concept developed earlier in this chapter and illustrated in Figure 4), who within Defense Systems Associates is responsible for interfacing with these individuals, and initiate a dialog between the two organizations.

One approach might be for all of the Defense Systems Associates personnel identified in Figure 3-4 to contact their Army counterparts, and start the conversation with a simple question:

"What are we doing wrong?"

Defense Systems Associates might also consider taking a hard look at itself, much as Tom Axelson and Aldo Pietras did in their initial conversation, and identifying what problems need to be addressed. The first step in solving any problem is to define the problem, and based on Defense Systems Associates' circumstances, it is obvious there are quite a few problems. For starters, Aldo Pietras might consider talking to his executive staff and all of the people in Defense Systems Associates to emphasize the fact that they are all there to work as a team to satisfy their customers, both internal and external.

Summary

This chapter developed the customer concept, and the idea that everyone has a customer. There are internal customers and external customers, and both sets of customers have needs and expectations that organizations and individuals committed to quality should strive to meet. Lee Iacocca once advertised that Chrysler has only three rules: Satisfy the customer, satisfy the customer, and satisfy the customer. From a TQM perspective, that's a good philosophy for running an organization.

References

"Soothing the Savage Customer," *Best of Business Quarterly*, Christopher W.L. Hart, James L. Heskett, and W. Earl Sasser, Jr., Winter 1990-91.

"Is Total Quality Management Failing in America?" *The Quality Observer*, Jim Clemmer, August 1992.

Marketing High Technology, William H. Davidow, The Free Press, 1986.

Managerial Engineering, Ryuji Fukuda, Productivity, Inc., 1983.

In Search of Excellence, Thomas J. Peters and Robert H. Waterman, Harper and Row, 1982.

Ballistic Systems Division Total Quality Management Handbook, Air Force Systems Command, Headquarters Ballistic Systems Division, October 1989.

Chapter 4

Quality Measurement Systems

Where it all begins...

George Cannelli was a proud man, and it showed as he toured the facility with Captain Ed Bowen, the Navy encryption program manager. At age 38, Cannelli had recently become president of PNB, a small electronics manufacturing organization with both military and commercial contracts. PNB specialized in circuit card assembly and related electronics integration. The company had several lucrative contracts. Captain Bowen was the Navy's program manager on one such program, a $40 million electronics production program.

Cannelli showed Captain Bowen PNB's wave soldering machine (a sophisticated mass production device for soldering electronic components to circuit cards) and the company's vapor degreasing machine (another complex piece of equipment for cleaning solder residue from newly assembled circuit cards). "You can see from our facility that we use only the latest production equipment," Cannelli said to the Captain, "and our production rate is excellent. We can produce in excess of 400 boards an hour."

Captain Bowen nodded his head understandingly. "That's a commendable production rate," the Navy officer said, but it was obvious his attention was elsewhere. Captain Bowen pointed to a group of workers soldering at the end of the room, partially obscured by the vapor degreasing machine. "What do those people do?" Bowen asked.

"They handle our rework," Cannelli explained. "Some of the boards that come off of the wave solderer have minor soldering defects. Whatever comes off the wave soldering machine less than perfect is made perfect by these people. They're the best solderers in the plant. We don't use them for anything else."

"I see," said Captain Bowen. "What kind of scrap and rework are you seeing here?"

"Very little," Cannelli answered quickly.

"How little?" the Captain asked. Cannelli didn't answer right away.

Captain Bowen continued. "How do these people know how well they're doing out here? I mean, other than the eight soldering specialists down there who do your very little rework on a fulltime basis, how do the rest of the people in the factory, or you for that matter, know how well you're really doing?"

Cannelli stared blankly for only an instant, and then took the Captain by the arm. "Let me get back to you on that," he said. "Let's go to our failure analysis laboratory. I want to show you our new scanning electron microscope. We paid over a hundred thousand dollars for it."

Cannelli wondered why the Navy officer seemed so interested in PNB's scrap and rework. After all, those people were only there to make sure that everything that went to the Navy met specification.

Measuring Quality

Quality is an abstract concept. Most people recognize quality when they see it, or the lack of it when they don't. How does one define quality, though? What does it really mean?

Think about the last time you purchased something like an appliance, or a camera, or anything even slightly complex. How did you feel when you opened the box? If you're like most of us, you were excited (especially if your purchase was for a discretionary item, like a new camera or a computer accessory). You probably felt a little bit like a kid opening a present, with all the attendant pleasures. But think about your other feelings. Did you harbor a subtle fear that the thing would not work? If your new purchase worked perfectly, all of its mating parts fit together well, and it did everything else you expected it to, what was your inevitable conclusion?

This is a thing of quality.

Think further, if you will, about the abstractness of the above statement. If an item meets one's expectations, the normal reaction is to judge it to be a thing of quality. If it doesn't, it is judged to be of low quality. And what about expectations? What are we really talking about here? Suppose you purchased a camera that took marvelous pictures, but its pieces fit together poorly and it generally exuded an aura of poor workmanship. Do you really have grounds for feeling uneasy? After all, the company that made the camera and the store that sold it more or less promised to deliver a camera that takes good pictures, and in our hypothetical situation, the camera does. So are you being fair in judging the camera to be of poor quality?

You bet you are. Why? Because people judge quality by how well the thing being judged meets expectations. That's an important point, because expectations frequently exceed the minimum standard suppliers are expected to meet. This point is so important that it will later be the subject of an entire chapter in this book.

With the above in mind, let us turn back to George Cannelli at PNB Electronics. His job is even more complex. His challenge is not to judge the quality of a single item purchased for personal use, but to instead assess the quality of one of the most complex and sophisticated things in the world: a manufacturing operation and the products it produces. How can George Cannelli evaluate the quality of his company? How does he know when it is good or bad, improving or getting worse, or acceptable or unacceptable?

Where Are We, and Where Are We Going?

In a previous chapter, quality management was defined as the process of continually improving an organization's products and services to better satisfy customer requirements and expectations. Two questions emerge from that definition:

- Are the customer's requirements and expectations being satisfied?

- Is the organization's compliance with customer requirements and expectations improving or deteriorating?

The above questions can only be answered through the implementation and use of a quality measurement system.

Hold that thought, and allow us to introduce yet another abstract concept by returning to George Cannelli. If George Cannelli purchased that hypothetical camera mentioned earlier, how do you think he would feel if the salesman who sold it to him called a few hours or days later and said: "George, I want you to come back to the store and pay us again for the camera." Would George be surprised and upset? If the answer is so obvious as to make the question insulting, consider this: Why isn't Cannelli as upset about those eight people at the end of the assembly line doing all the rework (the soldering specialists Captain Bowen took such an interest in)? Isn't Cannelli really paying those people to do the same thing he already paid others to do? Isn't he paying twice for the same thing?

Waste Is a Terrible Thing to Mind

The concept of undocumented customer expectations (such as the parts on the hypothetical camera discussed above fitting together well) will be developed in a later chapter. For now, the problem of measuring quality can be simplified by addressing just the known requirements. If you think this is an oversimplification, consider that many companies (and their employees) do not have a handle on questions as basic as these:

- How much product has to be scrapped because it does not meet dimensional or other requirements, and what is this scrap costing us?

- How much product has to be reworked or

repaired (either during the manufacturing process or after delivery to the customer), and how much is this costing us?

- What are the largest areas of scrap, rework, or repair?

- Which of the above items should we be working on fixing first?

- Are we working on any of the above?

- Do the people doing the work know what the scrap, rework, and repair rates are?

- Are we getting better or worse?

The bottom line is that many companies simply don't have a handle on their quality. They don't know how much of their effort is dedicated to building product versus how much is dedicated to reworking or repairing nonconforming product. They don't know how much is scrapped. They don't know how much poor quality is costing them, and they don't know whether things are getting better or worse.

If one were to ask the senior management of any company in America how much of their operation is being wasted due to scrap and rework, the answer one typically hears is quite similar to George Cannelli's: very little. We know, because we have asked the question many times, and with few exceptions the answer is either exactly as stated or so similar that the differences are meaningless. Some of the reasons for this ignorance will be cited shortly. For now, let us examine one of the best kept secrets in industrial America (a secret so subtle that even its keepers usually do not know it).

The Hidden Factory

During our development as an industrialized nation, American managers accepted the notion that inspection was the key to quality. The concept goes like this: One group of workers manufactures the product, and then a second group of workers inspects the output of the first group. The purpose of inspection is to sort the good product from the bad.

This concept has been inherent to industrial America for more than a century, but it is now hopelessly outdated, and any organization that attempts to operate in this manner is doomed to less than optimal profitability. Companies that rely exclusively or primarily on inspection will never realize their full profit or quality potential. This issue will be addressed more fully in a subsequent chapter on statistical process control, but for now, let us focus on companies that divorce the evaluation function from the production function. Let us consider its consequences.

Companies that rely on inspection are in the business of separating good product from bad. There is simply no other way to honestly explain the approach. These companies are detection oriented (as opposed to prevention oriented). They seek to detect nonconformances instead of preventing them. (A nonconformance is any deviation from requirements. This includes machined parts that do not meet dimensional requirements, discrepant material purchased from suppliers, finished assemblies that do not pass acceptance tests, items returned from customers, etc.)

This raises another key question: What do most companies do once they have detected product that does not meet requirements (or worse yet, what do they do when their customers detect these nonconformances)? Typically, such companies rely on the efforts of a third group of people whose efforts create the secret organization to which we referred earlier: the hidden factory. In addition to the normal production workers (the ones who build product the first time) and the inspectors (the ones who sort good product from bad), there is usually another group of workers dedicated solely to rework. We've been in plants where this number is as high as 25 percent of the total work force.

Sometimes the first group becomes the third (the workers spend significant portions of their time reworking or repairing product they previously built). Sometimes an entirely separate organization is used. Whether a company uses a separate group of workers or the ones who originally built the product, the point is that these workers take product culled out by the inspectors (i.e., the defective product) and rebuild it to meet requirements. These are the people who constitute the hidden factory.

The people themselves are not hidden, and it is usually obvious that they are working. What is not so obvious is that the reworkers constitute a second factory: the hidden factory. Their rework and repair activities are usually not isolated from the rest of the factory, and as a result, the rework (from both financial and productivity perspectives), becomes hidden.

Think back to the example at the beginning of this chapter, and recall PNB's solderers at the end of the production line (the workers who corrected the defects created earlier in the circuit card assembly process). Do you think George Cannelli has a handle on what their efforts are costing him? If Cannelli is like many managers, the answer is probably no.

How large and costly is the hidden factory? In our opinion, there are no reliable statistics for American industry, simply because most companies do not know what their rework costs are. But there are indicators. The U.S. Department of Defense Reliability Analysis Center found that poor quality costs comprise 15 percent to 50 percent of all business costs. A study by *USA Today* found that the cost of poor quality comprised 20 percent of gross sales for manufacturing organizations, and 30 percent of gross sales for service industries. When questioned on this subject, many chief executives guess their rework content to be below 5 percent. We find that figure to be woefully low. Many of the companies we have worked with were experiencing rework costs of 30 percent to 50 percent. Most companies are so accustomed to rework that they fail to recognize it.

At one high technology company (a producer of laser rangefinding equipment), the rework rate was effectively 100 percent. Every laser produced by the company was turned back somewhere in the process for readjustment, or replacement of failed components, or because it failed one of the in-process tests. Not a single assembly made it through the process, passed all tests, and was found acceptable for delivery the first time it went through the production process! This was a successful and profitable company (it essentially had no competition), yet think of what its profits could have been had it not been encumbered with such a heavy rework burden! Typically, the company didn't

believe it had a rework problem. It regarded its products as somewhat mystical, and its executives felt that the constant recycling was an inherent and unavoidable aspect of the business.

If you think the above problem is confined to ultra-high technology operations, think again. We have worked with several recreational vehicle manufacturers, who typically rework literally hundreds of defects on every vehicle before it leaves the factory.

So what's the bottom line? What does all of this mean? Here's what we believe:

- Quality is based on a product's compliance to expectations and requirements.

- Quality is a measurable characteristic.

- Quality measurement should be based on the quantity and costs of nonconformances.

- Poor quality raises costs unnecessarily, as poor quality increases the size and cost of the hidden factory (the scrap and rework content).

- Value and quality can be most efficiently improved by measuring nonconformances in terms of quantity and cost, and systematically attacking the dominant nonconformances.

The remainder of this chapter (and most of the rest of this book) describes how to implement management systems to accomplish the above.

Implementing Quality Measurement

With all of the above in mind, the challenge becomes defining, implementing, and using a quality measurement system to appropriately prioritize quality improvement actions. What quality data does a company need for this measurement and prioritization approach, and how does one going about getting it? The process and the data required are shown in Figure 4-1.

The process begins by identifying and recording nonconformances, collecting this data in a suitable data base, sorting the data from several different perspectives, and then using the data to efficiently drive the corrective action process.

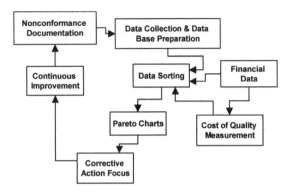

Figure 4-1. A Basic Quality Measurement System. This chapter provides a framework for capturing nonconformance data and costs, and structures a quality measurement system that drives continuous improvement.

Identifying and Recording Nonconformances

Quality measurement systems begin by identifying and recording nonconformances. This means that every time an inspector finds a nonconformance, every time an item fails a test, every time purchased parts are rejected, every time a statistically controlled process exceeds its control limits, and every time an item is returned from the customer because it failed to meet expectations, it is documented. Most companies use a document typically described as a Nonconformance Report, or a Nonconforming Material Report, or some other similar name (for our purposes in this chapter, we will call these Nonconforming Material Reports).

We recommend including the following information in Nonconforming Material Reports:

- A description of the item.

- A description of the nonconformance.

- The cause of the nonconformance.

- Disposition of the nonconforming item (typically, such items are either reworked or repaired to meet requirements, returned to the supplier if the item was purchased, or scrapped).

- Action taken to correct the nonconformance.

- The work area.

- The operator.

- The supervisor.

The identification and documentation of nonconformances may sound almost trivial, but we've found real problems in this area in many companies. Some of the companies with whom we've worked simply didn't have a system for recording the data. When nonconformances were discovered, they were simply sent on for rework or repair, or they were scrapped.

Most companies have a nonconformance reporting system, but in many organizations, there are subtle ways in which nonconformances escape being recorded:

- The inspector could simply give the part back to the operator for rework.

- The inspector could simply scrap the nonconforming item.

- The operator might rework the part without presenting it for inspection.

- The operator could simply scrap the part.

- Shop floor supervision could direct that the nonconforming item be reworked or scrapped.

- Discrepant purchased items could simply be returned to the supplier, without any documentation of the nonconformances.

Why this emphasis on documenting nonconformances? The answer is simple: We have to define a problem before it can be solved. Even the simplest item is subject to many nonconformances. These nonconformances have to be documented to allow for their identification, sorting from any of several perspectives (as we'll discuss shortly), and assignment to continuous improvement teams. If the nonconformances are not documented, the problems will tend to remain undefined, and continuous improvement will be elusive.

Documenting the nonconformance helps to force failure analysis and corrective action (most responsible people find it difficult to allow open documents to remain so for very long). Collecting data on nonconformances allows one to sort the data, rank order the nonconformances by frequency of occurrence or by cost, or by both. This rank ordering, as will be explained shortly, allows one to attack the biggest problems first (simply by revealing which problems are occurring most frequently, or are imposing the greatest cost).

Developing a Nonconformance Data Base

As nonconformances are recorded, one needs to collect the data in a manner that allows for rapidly determining nonconformances (by date, product, part, type of nonconformance, work area, operator, or supervisor), failure frequencies based on the preceding, trends in failure frequencies based on the above, and corrective action status for all nonconformances.

It's nearly impossible to do this job well without a computer. We've used Ashton-Tate's *dBase*, Microsoft's *Excel,* Microsoft's *Access*, and Lotus Development Corporation's *Lotus 1-2-3*; all have provided excellent service. These programs allow developing a data base that can provide the information described above based on examining the nonconformance data from several perspectives,[1] as will be described below.

The Elements of a Quality Data Base

The quality data base should include the information appearing on the Nonconforming Material Report (the source document for recording nonconformances). The concept is that the data base will allow for subsequent sorting, based on the criteria described above. One might be interested in learning, for example, which work area has the largest number of nonconformances. Or, perhaps it would be of benefit to know which parts or subassemblies are failing most often. If a design improvement is incorporated to improve reliability, knowing how many failures occurred before and after the design change would reveal if the upgrade corrected the problem, or if additional corrective action is required. A data base that includes the information discussed below can answer all of the above questions, as well as many others. Our recommendations for the elements of information to be included in a quality data base include the following:

- *Part Description*. This field should define the part by its part number and name.

- *Nonconformance Description*. This field should provide a simple description of the nonconformance. We recommend attempting to define what these nonconformances could be in advance, and providing standardized descriptions to either the inspectors or the data entry personnel. If you do the latter without doing the former, you have to make sure that your data entry personnel can accurately interpret the inspectors' comments and categorize these into the pre-defined "standard" categories. If the nonconformance description entries are standardized, this will allow for more accurate data base sorting later.

- *Date*. This field identifies the date the nonconformance was discovered.

- *Quantity*. This field identifies the number of components, subassemblies, or assemblies affected.

- *Nonconformance Cause*. The cause of the nonconformance should be explained in this field. In some cases, the cause will be immediately apparent. Examples include such factors as use of improper materials, incorrect assembly procedures, or obvious design deficiencies (as might occur if two mating parts meet their drawing requirements but do not fit together). In other cases, the cause of the nonconformance may not be immediately obvious, and further analysis will be required. One note of caution is in order here. Many

---

[1] For a sample data base file using Microsoft *Access*, with a downloadable file providing both input and output reports, please visit our website at www.bhusa.com

times, it is easy to blame a nonconformance on assembly technician error. These people will make mistakes, and when they occur, such mistakes should be noted. The reader should note, however, that in our experience, operator error constitutes only a small portion of the universe of nonconformance causes. Far more often than not, the underlying causes of apparent operator error include incomplete or inaccurate assembly instructions, poor lighting, poor design, inadequate tooling, or any of dozens of other factors. Do not hesitate to ask the technician what he or she suspects the cause of a nonconformance to be. As will be discussed in Chapters 5 and 6 (on problem solving and failure analysis), those closest to the work often have an intuitive feel for problem causes. Be aware of the fact that if you incorrectly attribute the cause of a nonconformance to be operator error, you'll most likely cut yourself off from this important source of information.

- *Nonconforming Item Disposition*. This field should include information describing the disposition of the nonconforming material. Typically, nonconforming items are either reworked or repaired to meet requirements, returned to the supplier (if the item was purchased), or scrapped.

- *Corrective Action*. This field describes the action taken to correct the nonconformance. We recommend providing a standardized list of options (or perhaps codes) to simplify the data retrieval task. We'd like to point out here that you should recognize that corrective action is entirely different than nonconforming item disposition. Disposition (as used here) simply describes what was done with the nonconforming hardware; corrective action describes what is being done to prevent recurrences of the nonconformance.

- *Work Area*. This field identifies the work area that created the nonconformance (e.g., machine shop, welding, accounting, paint department, etc.).

- *Operator*. This field lists the name of the assembly technician (i.e., the person who created the nonconformance). Some delicacy is required in developing this information. Obviously, the purpose here is to isolate the root cause of the nonconformance, and in this case, the information is needed to determine if specific operators associated with recurring nonconformances are adequately trained, if the work instructions are adequate, etc. Gathering such information serves two purposes. It allows making the above determinations. It also lets those performing and, as will be mentioned below, supervising the work know that their performance is being monitored.

- *Supervisor*. This field lists the name of the work area supervisor, for all of the reasons described immediately above.

- *Nonconforming Material Report Number*. We recommend numbering each Nonconforming Material Report. The document's number should be included in a data base field. If the Nonconforming Material Reports are stored in an organized manner, one can use the data base to sort by any number of attributes, identify the specific Nonconforming Material Report numbers associated with the attribute, and then retrieve the report for additional information or analysis.

A Basic Quality Measurement System

What can one do with the above data? The essentials will be presented shortly, but before delving into the specifics, let's consider the manner in which the data should be presented.

Our experience has confirmed that tabular data (i.e., rows of numbers) is too dry for communicating critical quality measurement data to business leaders, middle managers, and workers. If tabular data is used for communicating quality measurement information, the people who should review and understand the data may not (this includes those doing the work, and others empowered to take actions to lower nonconformance rates). We have, however, obtained excellent results (and observed others do the same) using Pareto charts.

Pareto was an Italian economist (he lived from 1848 to 1923) who developed what was to become known as the 80/20 rule. This rule holds that in any situation, 80 percent of the results are typically attributable to 20 percent of the causes. Pareto initiated this finding by observing that the wealth of most nations is concentrated in only a small portion of the population.

Others were quick to seize upon Pareto's concept and apply it to other fields. J.M. Juran, one of the guiding lights in modern quality management philosophy, applied Pareto's name to a format for identifying the "vital few and trivial many" quality defects (a phrase created by Juran).

The Pareto charting concept involves developing bar charts to show the quantity or cost of each defect category, with the largest items appearing first, followed in descending order by the second largest, third largest, and so on. This approach allows one to rapidly separate the most significant quality issues from those that are trivial. Several Pareto charts will be illustrated in this chapter.

We recommend using Pareto charts extensively (but not exclusively) for conveying quality measurement results. We will also recommend other graphical quality measurement formats, including trend lines and pie charts. All serve but two purposes: to help an organization rapidly understand where its largest problems lie, and if things are getting better or worse.

Here are the reports we recommend culling from the quality data base to form a working quality measurement system:

- Summaries of nonconformance quantities and costs.

- Summaries of scrap, rework, and repair actions.

- Summaries of supplier performance.

- Summaries of product reliability (i.e., how the product performs after delivery to the customer).

- Summaries of what quality is costing the

organization.

The quality measurement presentations we recommend are summarized in Figure 4-2, and described in detail in the following pages.

Figure 4-2. A Summary of Recommended Quality Measurement Presentations. We recommend Pareto, Trend, and Pie Chart presentations in each category, as will be illustrated below.

Nonconformance Summaries

The first set of data we recommend presenting consists of a set of three nonconformance summaries:

- A simple bar chart showing the types and quantities of nonconformances and the number of Nonconforming Material Reports over time (this is a trend chart).

- A set of Pareto charts showing the above data organized along program or product lines for each reporting period.

- A set of Pareto charts showing the program or product line data expressed in terms of cost.

The first nonconformance summary shows both the quantity of nonconforming parts and the number of Nonconforming Material Reports over time, as shown in Figure 4-3 (we'll stick with PNB Electronics to illustrate the concepts developed here).

We recommend showing data for the entire company in the nonconformance summary trend chart, as this summary chart provides a quick overall indication of the company's quality direction. It's either going up, or going down, or drifting aimlessly. As Figure 4-3 shows, PNB Electronics' quality is improving (notwithstanding George Cannelli's lack of information concerning his scrap and rework costs).

PNB Electronics
Nonconformance Summary

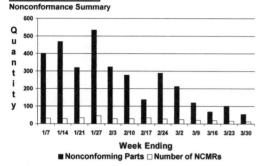

Week Ending
■ Nonconforming Parts □ Number of NCMRs

Figure 4-3. PNB Electronics' Plant-Wide Nonconformance Summary Chart. This chart shows the number of nonconforming parts and the number of Nonconforming Material Reports for each week. PNB's quality is improving, as shown by the downward trending number of nonconformances.

Showing both the quantity of nonconformances and the number of Nonconforming Material Reports is important, as it quickly reveals if the organization is suffering from multiple nonconformances. Stated differently, if the number of nonconforming parts far exceeds the number of Nonconforming Material Reports, it is showing that some of the nonconformances are occurring in large numbers before being discovered. This should steer the corrective action effort, as it has implications bearing on inspection points, statistical process control implementation, operator training, etc. (all of these concepts will be covered in subsequent chapters). This report is also useful for showing trends in total number of nonconformances, which provides a good indication of an organization's success in attaining quality improvements.

We recommend Pareto charts for the second nonconformance summary category. This category of information should be organized along program or product lines, and show the number of nonconformances in relative order of occurrence.

A sample Pareto chart for one PNB Electronics

program, the Navy encryption device mentioned at the beginning of this chapter, is shown below in Figure 4-4. As Figure 4-4 shows, the most dominant nonconformances include resistor failures, soldering defects, circuit card delamination, and incorrect component installation. The information shown in Figure 4-4 shows the management of PNB Electronics that these are the areas in which they should consider applying quality improvement activities. Figure 4-4 also shows that within the Pareto chart, data can be overlain for several months, thereby showing not only the relative frequencies of occurrence for each nonconformance, but also whether the trend for each nonconformance is improving or deteriorating.

Navy Encryption Program
PNB Electronics Monthly Nonconformances

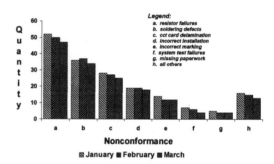

Nonconformance
▨ January ▧ February ■ March

Figure 4-4. Nonconformance Quantity Pareto Chart for PNB Electronics' Navy Encryption Program. The chart shows the dominant nonconformances in descending order for the last three months, based on quantity of nonconformances.

Navy Encryption Program
PNB Electronics Monthly Nonconformances

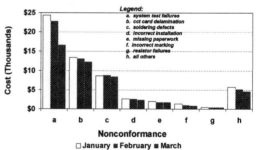

Nonconformance
□ January ▨ February ■ March

Figure 4-5. Cost-Based Pareto Chart. This chart converts the nonconformances shown in Figure 4 to costs. Note that the highest quantity nonconformance in Figure 4 (the failed resistors) is low in cost and does not appear as a dominant quality detractor in this chart, while others that were low in Figure 4 are dominant quality detractors by virtue of their cost.

Figure 4-5 shows a Pareto chart that presents the data described above from another perspective. This chart takes the nonconformance quantity data shown in Figure 4-4 and converts it to cost.

The usefulness of a cost-based Pareto chart is that it can alert management to the low quantity nonconformances that impose high cost. In the examples included here, the highest count nonconformance is associated with PNB's open resistors (see Figure 4-4), but these items only cost about $10 each. The 47 nonconforming resistors cost PNB $470. In the month of March, PNB experienced just four system failures in acceptance testing, but the cost of troubleshooting these failures and incorporating corrective action exceeded $16,700.

The message here is simple: Nonconformance quantity is but one quality measurement index. Costs are also significant in determining where corrective action should be focused. Through the use of these Pareto charts, the implications for corrective action are straightforward. Clearly, PNB's quality improvement efforts should be focused on improvements to the system design and assembly process to eliminate the system failures that were discovered during system acceptance testing. Without a cost-based Pareto chart, the requirement to focus on system performance issues might have been obscured by the high quantity of failed resistors and other items.

The cost issues should not be the only factor guiding PNB's continuous improvement efforts, however. Those high quantities of failed resistors also require attention. Both the quantity- and cost-based Pareto charts should guide the focus of PNB's corrective actions, such that the most significant quality problems are systematically identified and attacked.

Our experience leads us to believe that the nonconformance Pareto charts described above should be prepared on a monthly basis, as the week-to-week fluctuations inherent to any organization could mislead the corrective action effort. You should consider if this is the case in your organization, and update the charts recommended herein on a frequency most useful to your organization. It's conceivable that on highly dynamic programs it may be necessary to update the

charts weekly, or perhaps even daily. On more stable programs, it may be beneficial to update the charts on a quarterly basis.

There's one additional set of nonconformance charts we recommend to provide quick management visibility into an organization's quality challenges. These are pie charts that categorize the organization's total nonconformances by both program (or product line) and work area.

The program-based approach shown in Figure 4-6 should alert George Cannelli to the fact that most of his quality problems are coming from the Navy encryption device program.

Program Nonconformance Summary
PNB Electronics - March 1996

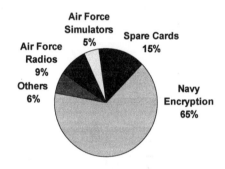

Figure 4-6. Program-Based Nonconformance Pie Chart. This chart summarizes nonconformances for the entire company.

The work-center pie chart shown in Figure 4-7 should similarly show Cannelli that most of the nonconformances emanate from the circuit card assembly area. This should tell him to assign people to reduce the circuit card defect rate.

The nonconformance summaries described here serve several important functions. The summaries provide immediate insight into the most frequently occurring nonconformances and the costs of these nonconformances (from both a company and a program or product line perspective). The summaries rapidly reveal if the organization is succeeding in its quest for continuous improvement. Perhaps most significantly, the summaries provide direction for the organization's quality improvement efforts.

Work Center Nonconformance Summary
PNB Electronics - March 1996

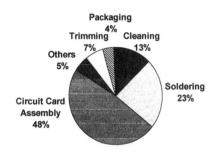

Figure 4-7. Work-Center-Based Nonconformance Pie Chart. This chart summarizes nonconformances, by work area, for the entire company.

Scrap, Rework, and Repair Summaries

Scrap is material that is unacceptable for further processing or delivery to the customer, either because it cannot be reworked or repaired to meet requirements, or because doing so costs more than simply throwing the item away and replacing it. How should a company record and report its scrap? We recommend collecting and presenting scrap data on at least a monthly basis. Our experience shows that two sets of data work well: a Pareto chart showing items scrapped by quantity, and a Pareto chart showing items scrapped by cost.

The first Pareto chart simply presents a summary of the total quantity of each item scrapped. The second set of data takes the first set of data (i.e., the total quantities of each item scrapped) and converts it to dollars in much the same manner as described for the nonconformance summaries above.

Rework is defined as any work performed to make nonconforming product comply with drawing and specification requirements. We recommend that rework data be presented monthly, in the same manner described for scrap reporting: a Pareto chart showing the types and quantities of items reworked, and a Pareto chart showing rework costs. The first and second Pareto charts provide the same kinds of data (and the same usefulness in guiding corrective action) as do the Pareto charts discussed earlier for nonconforming parts quantities and costs. Corrective action should be focused on those items

with the highest rework content or cost.

Within the aerospace and defense industry, repair is distinctly different than rework (a more detailed explanation is included in Chapter 16). While rework describes actions required to bring an item into compliance with drawings and specifications, repair refers to work that will bring an item into compliance with performance requirements, but not the engineering drawings and other data describing its configuration. A good example of a repair is welding two broken pieces together. The item will meet its performance requirements, but the original design does not include a weld. We recommend presenting repair data separately from rework data, but the reporting format should be the same (Pareto charts showing repair quantities and costs).

Supplier Performance Summaries

Most manufacturing companies today typically buy a high percentage of their products from suppliers. We'll review supplier quality issues in Chapter 15, but let's first consider tracking supplier quality performance with a quality measurement system.

It makes sense to keep track of suppliers' quality performance to guide the award of future contracts based on supplier quality history, which drives the true cost of doing business with a supplier. A supplier may win a contract by submitting the lowest bid, but if the supplier's products have a high rejection rate or are habitually delivered late, the significance and attractiveness of the original low bid fade. The true cost of doing business with such a supplier is much higher than the initial low bid. The true cost is difficult to quantify, but it includes the costs of late deliveries from your company (if your products are delivered late as the result of supplier delinquencies), the administrative and other costs that result from defective supplier product, other interruptions to your organization's schedules, etc.

We recommend developing a tracking system for all of your suppliers to record delivery delinquency rates, product rejection rates, and some or all of the factors mentioned above. You may wish to consider developing a weighing scheme tailored to the factors most important to your organization. Some companies consider defect rate to be of paramount importance and weigh it heavily in their supplier rating scheme. Others weigh schedule compliance more heavily. Most companies place a lighter emphasis on paperwork accuracy (although we would be the first to admit that paperwork deficiencies can interrupt production as quickly as defective product or delinquent deliveries). The concept is that weighing the various supplier quality factors meets several important needs. These needs include forcing a definition of those factors most important to your organization, structuring a supplier rating system that provides meaningful quality measurement indices to guide the selection of suppliers best matched to your needs, and making your suppliers aware of what's important to you.

To gather information for use in the supplier rating system, we've found that the Nonconforming Material Report identified earlier in this chapter works well. The nonconformances on these forms should be identified as supplier-related deficiencies to prevent artificially inflating the nonconformance summaries for your company's internal operations.

As mentioned above, most companies with supplier rating systems tailor their own criteria. In addition to tailoring a supplier rating scheme, we also recommend specifying the minimum quality requirements suppliers must meet if they wish to be kept on an organization's approved supplier list. Establishing supplier quality criteria in advance (and in concert with your company's procurement organization) is usually a good idea, as it eliminates much of the subjectivity that would otherwise be associated with a decision to cease doing business with low quality suppliers.

Let's consider an example George Cannelli might use at PNB Electronics. We'll assume that Cannelli accepts the supplier rating criteria we typically recommend as a starting point: Suppliers delinquent on more than 10 percent of deliveries or with defect rates exceeding 2 percent should not be used (regardless of the weighing concept), and suppliers who deliver on schedule with zero defects should receive a score of 100 percent. Suppose PNB decides that defect rate is four times as significant as schedule compliance. Let's further suppose that after discussion with his procurement manager, Cannelli decides that every percentage point in missed deliveries results in a lost point. Cannelli

wants to score suppliers' quality performance similarly, except he wants to weigh the results by a factor of four. Every percentage point in defect rate would therefore result in a four point loss. Finally, Cannelli and his procurement manager want to add one additional qualifier. Along with the criteria described above, a minimum weighted score of 95 will be required for suppliers to continue to do business with PNB. Based on the above criteria, here's how PNB would rate suppliers:

- Those with delinquencies exceeding 10 percent or defect rates exceeding 2 percent would be excluded from further consideration.

- Those with delinquencies at 10 percent or below, and whose defect rates are at 2 percent or below, would have a score calculated as follows:

$$SRS = 100 - 4D_e - D_l$$

where:

SRS = supplier rating score
D_e = defect rate
D_l = delinquency rate

To continue our example, suppose PNB is considering five resistor suppliers: Quantel Technology, Orion, Hexagon Products, Faritsu, and DataPoint. Quantel Technology was delinquent on 11 percent of its deliveries to PNB last year; therefore, they are excluded from further consideration. Orion had no defects, but they were late on 3 percent of their deliveries (for a score of $100 - 4*0 - 3 = 97$). Hexagon Products met all schedules, but had a 2 percent defect rate (for a score of $100 - 4*2 - 0 = 92$). Faritsu was late on 1 percent of its deliveries, and had a defect rate of 3 percent (for a score of $100 - 4*3 - 1 = 87$). DataPoint was late on 2 percent of its deliveries, and had a 0.5 percent defect rate (for a score of $100 - 4*.5 - 2 = 96$).

How should the above supplier rating data be presented? Once a company has tailored a weighing scheme (as described above), we recommend comparing the weighted quantitative score for each supplier against a minimally acceptable standard.

We've found a simple bar chart to be effective for these presentations, as shown in Figure 4-8.

Regardless of the scoring criteria or the presentation approach, the point here is simple: You need to know a supplier's quality performance track record to make informed decisions about continuing a business relationship. You can't make informed decisions without developing and maintaining a supplier quality data base.

PNB Electronics
Resistor Supplier Performance

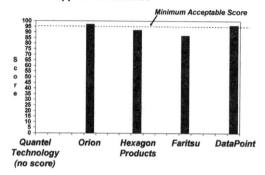

Figure 4-8. Supplier Ratings for PNB Resistor Suppliers. The above chart suggests that Orion should be selected based on quality criteria, while Quantel Technology, Hexagon, and Faritsu should be excluded from further consideration.

As mentioned at the beginning of this section, most companies today procure a substantial portion of their product content from suppliers. It makes sense to collect, evaluate, and base selection decisions on supplier quality data.

Product Reliability Summaries

Knowing how the product or service provided by your company performs after delivery to the customer is of paramount performance. This data may be difficult to obtain, but it provides perhaps the most direct indication of customer's perceptions of your quality. Quality is the degree to which customer requirements and expectations are satisfied, so knowing what the product is doing after delivery is critical to any quality measurement system. Obtaining this data (and working to reduce the most prevalent nonconformances) will provide a high return for two reasons: Customers will recognize your quest for continuous quality improvement, and post-delivery nonconformances are the most expensive to correct, so anything done to identify and eliminate them prior to delivery will

greatly reduce your organization's cost of quality.

In many cases, obtaining post-delivery product reliability data will be challenging. The difficulties associated with obtaining such data can be overcome by tapping three sources of information: warranty data, repair data, and other direct inputs from the customer. Your company or its repair stations should be able to provide warranty and other repair data. If the information is not available, we recommend that you make arrangements with the repair centers to collect it. We've seen simplified versions of the Nonconforming Material Report described earlier work well.

We also recommend approaching customers directly for additional information on product reliability. If you only have a small number of customers (for example, if you sell to the military), your customers would probably be happy to provide the data. If you sell to many customers (perhaps your company produces a consumer product), statistically sampling customers may make sense. Regardless of the approach, your company's reputation for quality will improve simply by asking for post-delivery product reliability data. Customers will recognize your need for the data, and your desire to better meet their requirements and expectations.

We recommend the Pareto chart format for this data. An approach that works well is to present monthly post-delivery failures in a Pareto chart that shows the results for three months (as illustrated earlier in Figure 4-4).

Cost of Quality Summaries

The cost of quality is yet another important quality measurement index. Quality costs will be defined in the following paragraphs, but for now, consider this radical thought: The cost of quality in a perfect organization should be zero. Why? In an ideal environment, everyone would do his or her job perfectly. In an ideal company, the design would contain no flaws (the engineers did their jobs perfectly), the materials and purchased parts used in building the product would meet all requirements (all of the suppliers performed perfectly), and the product would be perfectly manufactured (the factory technicians and assembly personnel would do their work without error). In such an environment, would there be a need for any inspectors? Or technicians to perform rework? Or engineers to adjust the design as design flaws emerge in production? The obvious answer, in our idealized and perfect environment, is no. None of the above activities would be needed. As a *goal*, there should be zero expenditures for quality.

As a practical matter, we live in an imperfect world. It would be naive to assume that any company could afford to spend nothing on quality. Can the last sentence be turned around, though? If the goal is to provide a quality product, can one expect to make the costs of quality approach zero? Our experience confirms that within limits, the answer is yes. That's good news, because companies become more profitable and consumers receive better goods and services. (The practical limits are driven by costs associated with preventive quality measures intended to curtail larger detection-based costs, as will be explained below.)

Given all of the above, how should one capture and measure the cost of quality? Many of the concepts have been mentioned already in the sections in which we discussed the preparation of Pareto charts based on not only the quantities, but also costs of scrap, rework, repair, and other nonconformances (see, for example, Figure 4-5). In addition to these quality cost indices, many organizations capture, segregate, and track quality costs in three areas: failure costs, appraisal costs, and preventive costs. Each category is explained below.

- *Failure Costs*. Failure costs are those associated with correcting nonconforming material, including scrap, rework, repair, warranty actions, and others related to the correction of nonconformances. Many organizations further subdivide this category into internal and external failure costs. One reason for doing so is that internal failure costs are a measure of a company's operating efficiencies, while external failure costs provide measures of both product quality and customer satisfaction. Ideally, failure costs should approach zero. Failure costs typically comprise between 70 and 85 percent of an organization's total quality costs.

- *Appraisal Costs*. Appraisal costs are those related to the detection of defects. This cost category includes the costs of inspection, testing, and other measures used to separate good product from bad. Failure analysis and other activities focused on identifying underlying nonconformance causes should also be included in this category. Our experience (as well as that of others) shows that appraisal costs average around 15 percent of an organization's quality budget.

- *Preventive Costs*. Preventive costs are those associated with activities designed to prevent defects. This is the area one would hope to have dominate an organization's quality budget. Such costs include participation in the design process to eliminate potential failure modes, process improvements designed to prevent production of nonconforming hardware, generation of Quality Function Deployment data (this concept will be addressed in Chapter 11), and others. Unfortunately, most organizations' preventive costs are relatively insignificant when compared to the failure and appraisal costs. The intent of any organization should be to lower failure and appraisal costs as a result of an intelligent investment in prevention-oriented activities. As will be covered elsewhere in this book, the costs of preventing nonconformances are trivial compared to the costs of detecting or correcting them.

We recommend presenting cost of quality data on a monthly basis in four formats. The first is a pie chart to show the organization's monthly failure, appraisal, and preventive cost (an example for PNB Electronics is shown in Figure 4-9). This chart is somewhat redundant to the other charts we'll describe below, but we believe it's useful because it provides a simple and quick portrayal of an organization's quality cost structure.

Figure 4-10 shows quality costs in each of above categories over time. Figure 4-11 shows total monthly quality costs as percentages of other costs over time, which further helps to put quality costs in perspective. These relative displays (as shown in Figure 4-11) typically should include total quality costs as a percentage of total sales, rework and repair costs as a percentage of total labor costs, and scrap costs as a percentage of total material costs.

Cost Of Quality Makeup
PNB Electronics - March 1996

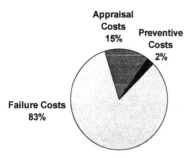

Figure 4-9. PNB Electronics' Monthly Cost of Quality Makeup. This presentation offers a quick look at how the organization is investing its quality dollars.

In addition to showing the relative makeup of quality costs (as seen in Figure 4-9), it is a good idea to show quality cost trends, as shown by Figures 4-10 and 4-11.

We've heard people say that cost of quality industry averages in the manufacturing sector average around 30 percent, but little hard data is available. (Such data is competition sensitive, and many companies do a poor job capturing and recording quality data).

A month-to-month review of the trend charts such as those shown in Figures 4-10 and 4-11 will indicate if the organization's cost of quality is increasing or decreasing. This provides indications to guide the implementation of continuous improvement efforts.

These indicators are quite helpful, as they allow management to visualize the comparative size of the hidden factory mentioned earlier in this chapter, and to determine if the costs of the hidden factory are moving in the right direction.

There's one other cost of quality chart we recommend, and it's a monthly summary Pareto chart showing the dominant costs in each of the quality cost categories. This a good idea, as it allows one to rapidly determine the drivers in each quality cost category, and therefore, where it makes sense to apply actions to reduce these costs. The concept is illustrated in Figure 4-12. Again, the thought is to appropriately focus continuous improvement efforts.

Cost of Quality Trends
PNB Electronics - Overall Costs

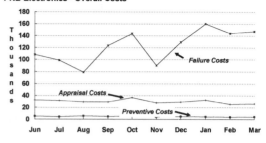

Figure 4-10. PNB Electronics' Cost of Quality Trends. This chart shows preventive costs to confirm the results shown in Figure 4-9 and the other trend lines here. PNB spends very little on preventive quality concepts, which has resulted in high appraisal and failure costs.

Cost of Quality Relative Trends
PNB Electronics

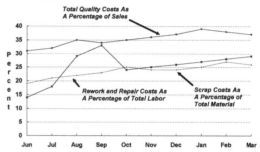

Figure 4-11. PNB Electronics' Relative Quality Cost Trends. This presentation helps to put quality costs in perspective. PNB's quality costs exceed 30 percent of sales! Scrap and rework are also significant compared to normal labor and material costs. Reductions in these costs would flow straight to PNB's bottom line as increased profit!

Cost of Quality Pareto Chart
PNB Electronics - March 1996

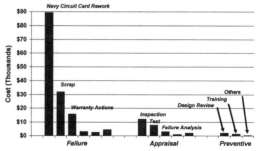

Figure 4-12. Cost of Quality Pareto Chart. This chart rank orders the contributors in each of the quality cost categories: failure costs, appraisal costs, and preventive costs.

Quality Issue	Required Corrective Actions	Assigned To	Required Completion Date	Status
High rework content on Navy Encryption Device circuit cards	Identify rework categories	Smith	13 May 1993	Complete
	Prepare fault tree analyses to assess root causes	Statler	30 May 1993	Under way
	Implement required corrective actions	Randolph	1 July 1993	Awaiting analysis results
Improperly completed NCMRs	Document process using flow chart approach	Goren	5 May 1993	Complete
	Rewrite procedure	Goren	7 May 1993	Complete
	Conduct training	Moffet	9 May 1993	Under way

Figure 4-13. Excerpt From Quality Improvement Action List. We recommend preparing such lists based on the indicators provided by an organization's quality measurement system, and including a description of the issue, the required corrective actions, the personnel responsible for executing each action, required completion dates, and a status summary.

Forcing Corrective Action

The principal objective of a quality measurement system is to appropriately focus corrective action and other continuous improvement efforts. All of the data in the world would be of no use if nothing were done with it. The challenge one faces in structuring a quality measurement system is to present the data in a manner that facilitates absorption and action by those to whom it is targeted. Simply stated, you want the right people to read and understand what the quality measurement system reveals, and then to act on the information such that the causes and costs of poor quality are continuously reduced.

To maximize the effect of the charting and presentation techniques discussed in this chapter, we recommend regularly presenting quality measurement data to those empowered to do something about it: the organization's senior and middle management, and the people actually building the product or delivering services. We've found four techniques effective for providing this visibility and driving continuous quality improvement. Each is described below.

- *Quality Measurement Reports*. Experience has convinced us that providing two quality measurement reports (one weekly, the other monthly) helps to maximize the effectiveness of quality measurement data. The first report, the weekly quality summary, shows the total number of nonconformances and the cost of these nonconformances, and a brief description of the nonconformances. A weekly quality newsletter is a particularly effective technique for providing this data. The monthly report should include the more detailed quality data described throughout this chapter, including program/product quality summaries (based on both nonconforming item quantities and cost), summaries of scrap, rework, and repair actions, a summary of supplier performance, a summary of product reliability (including warranty actions), and a summary of what quality is costing the organization (as described above). We recommend distributing this summary to your organization's executive staff, and others who can implement actions to correct the dominant quality detractors.

- *Quality Status Presentations*. Progressive organizations recognize that quality status is a critical success factor. Accordingly, many organizations are making quality status an integral portion of all management reviews. We recommend tailoring the information to the review. For program or product specific reviews, it's appropriate to include only the quality measurement indices related to that program. For company-level meetings, it makes sense to include such overall quality measurement indices as company-wide costs of quality, dominant nonconformances in each program, and others. The content of these presentations varies from company to company. You have to judge what's appropriate for your organization. There's one other factor we'd like to mention here: When you present the dominant quality deficiencies, you should also present a status of the actions being pursued to correct each nonconformance. If no actions are under way, we suggest that you present the actions you recommend. The managers (and others) to whom the quality measurement information is targeted need to know that the problems are being worked on, or what needs to be done so that the continuous improvement effort retains its momentum.

- *Work Area Quality Data.* In many cases, the people on the factory floor or those who are actually providing services to your customers are the ones who can do the most to improve quality. For that reason, it makes sense to provide work-area-specific quality data. We recommend prominently displaying Pareto charts in each work area (or the office area of service industry members) to show the nonconformances for that work area. We also recommend annotating the top three or four items on these Pareto charts to describe the corrective actions under way. This shows employees where the quality challenges lie, and it shows that management is interested in sustaining continuous improvement.

- *Quality Improvement Action Lists.* To help close the corrective action loop, we recommend summarizing dominant nonconformances and other quality challenges identified by the quality measurement system, the corrective actions required to eliminate the root causes of the nonconformances, the personnel to whom these actions have been assigned, and required completion dates. Figure 4-13 on the previous page shows the way we've structured such lists. Quality Improvement Action Lists keep things moving. They prevent required corrective actions or other continuous improvement efforts from "falling through the crack." Our advice is to publish such lists monthly.

Summary

This chapter began by examining the challenges faced by George Cannelli at PNB Electronics. Cannelli's challenges include getting his arms around his hidden factory, determining what's driving the scrap, rework, and repair going on in his plant, and implementing the actions necessary to eliminate these unnecessary cost drivers and quality detractors. The only way to do this is to systematically identify all of the nonconformances, record these nonconformances in a quality data base, and then rank order the nonconformances from both quantity and cost perspectives to arrive at the most logical targets for continuous improvement activities.

This chapter provided a framework for meeting the first two challenges: identifying and ranking the nonconformances. Implementing continuous improvement in an appropriately focused manner mandates understanding where the dominant quality detractors lie. Without a structured quality measurement system, this simply is not possible.

This chapter provided a framework for an effective quality measurement system, which is where effective TQM and continuous improvement begin. The next question is: Once the nonconformances are known and rank ordered, how can they best be eliminated? The remaining chapters of this book develop the concepts and technologies needed to answer this question.

References

"Cutting Quality Costs," Asher Israeli and Bradly Fisher, *Quality Progress*, January 1991.

"In Search of Six Sigma: 99.997% Defect Free," Brian M. Cook, *Industry Week*, 1 October 1990.

"Improved U.S. Defense Total Quality Control," Thomas R. Stuelpnagel, *National Defense*, May/June 1988.

"How to Manage Quality Improvement," Charles C. Harwood and Gerald R. Pieters, *Quality Progress*, March 1990.

Cost of Quality, Defense Logistics Agency Special

Training Course, February 1987.

MIL-Q-9858A, "Quality Program Requirements," Section 3.6, Department of Defense, March 1985.

Guide to Quality Control, Kaoru Ishikawa, Asian Productivity Organization, Nordica International Limited, 1982.

How to Use and Misuse Statistics, Gregory A. Kimble, Prentice-Hall, Inc., 1978.

Chapter 5

Problem Solving

A simple four-step problem-solving process that works...

John Lavery had an enormous problem in front of him. As a senior engineering manager at General Digital Electronic Systems, he had participated in problem-solving exercises and failure analyses many times in his career. None of the problems he had ever worked on previously was as significant, however. The United States was about to go to war in the Persian Gulf, and the system to which Lavery was assigned, the Suspended Laser Acquisition Pod, did not work.

This was the real thing. It wasn't as if the system would simply fail an acceptance test and the company could rework the hardware. The repercussions weren't as simple as a delay in shipments with the consequent delays in payment. Saddam Hussein had thousands of tanks massed in Kuwait along the Saudi border, and U.S. troops were preparing to go into combat against them. If the Suspended Laser Acquisition Pod did not work, smart munitions would miss their targets, and Allied troops would die.

Background

General Digital Electronic Systems developed and manufactured the Suspended Laser Acquisition Pod (or SLAP, for short). The SLAP is a helicopter-borne surveillance, target acquisition, and target designation system. The system consists of a sensor suite that includes a thermal imaging sensor, a television sensor, a laser rangefinder and target designator, and associated environmental control, vibration-isolation, and other subsystems to support the sensors. The system is mounted in a spherical housing situated above the host helicopter's main rotors, as shown in Figure 5-1.

When the United States deployed troops in the Persian Gulf as part of the Desert Shield

preparations for war, General Digital had already built seven production lots of Suspended Laser Acquisition Pods. Hundreds of SLAP systems had already been delivered to the U.S. Army.

Figure 5-1. Suspended Laser Acquisition Pod. The SLAP system is housed in the aerodynamic shell below the helicopter fuselage.

Electro-optical target acquisition and designation systems such as the SLAP have proven to be an invaluable asset to modern tactical air and ground forces. These systems provide passive target acquisition capabilities, which allow for platform survivability against anti-radiation missiles and other countermeasures. The laser rangefinding and target designation features of modern electro-optical systems provide extremely accurate ranging information and target designation capabilities to support a variety of smart munitions, and they greatly enhance the mission effectiveness of U.S. military forces.

SLAP Operation

In operational use, helicopter pilots use the SLAP television sensor to detect targets during daylight conditions. The thermal imaging sensor is similarly used to detect targets during night operations (the thermal imaging sensor relies on the targets' infrared heat signatures to produce an image for display to the pilot). Once a target is detected, the pilot can energize the SLAP laser to determine the range to the target, or to "paint" a laser spot on the target to provide a homing point for laser-guided munitions. The munitions, fired by another vehicle (either on the ground or in the air), then homed in on the reflected laser energy.

When Iraq invaded Kuwait in August of 1990, the United States deployed a significant military force to the Persian Gulf. Key elements of the Desert Shield deployment included the SLAP-equipped fleet of OH-58 helicopters. Prior to deploying overseas, the U.S. Army evaluated the combat readiness of its complex weapons systems. One such evaluation involved the Suspended Laser Acquisition Pod, and that's when John Lavery and General Digital learned that the system was not performing as intended.

During the pre-Persian-Gulf deployment SLAP system evaluation at Yuma Proving Grounds in southern Arizona, a disturbing finding emerged. The SLAP line of sight did not track with the laser line of sight, which resulted in a designation pointing error. The difference was significant (the SLAP designation pointing error exceeded specification limits). The nonconformance had a very real combat meaning: The deviation from requirements was enough to induce a miss for laser guided munitions. The result would, at best, mean degraded combat readiness. At worst, it could allow the target to defeat the launch platform. Clearly, the challenge was to solve the SLAP designation pointing accuracy problem, and to recommend appropriate corrective action.

The SLAP Failure Analysis

General Digital quickly assembled a team of top technical experts to identify the cause of the designation pointing error anomaly. The failure analysis team initially focused on the SLAP boresighting system, as the team intuitively felt that a designation pointing error would most probably be related to a deficiency in the boresighting system. The boresighting system is used to align the SLAP laser, thermal imaging sensor, and television sensor to each other, and to the system line of sight.

After concentrating on the boresighting system for

an extended period without finding the cause of the designation pointing error (although other system problems were uncovered and corrected), the problem-solving team focused their efforts elsewhere. Detailed investigations into issues related to the SLAP power supply and the system microprocessor were pursued; however, neither yielded the cause of the designation pointing error.

One of the problems faced by the failure analysis team included evaluating reams of technical and test data. Several months into the analysis, the problem-solving team recognized that the system test data showed that the laser rangefinder/designator and the television sensors met their designation pointing error specification requirements. Only the thermal imaging sensor exceeded its designation pointing error requirements.

SLAP Optical Train

The SLAP sensors and target designation features communicate optically with the outside world through two windows, as shown in Figure 5-2. The first window is the thermal imaging sensor window, which is made of germanium. Germanium is an opaque black material (germanium is transparent to the thermal imaging sensor's infrared wavelength, however). The laser and television sensors see through the second window, which is a clear aperture (the laser and the television sensors share many other elements of the SLAP optical train).

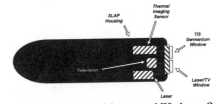

Figure 5-2. SLAP Internal Layout and Windows. Only the laser, television, and TIS are shown (other components are omitted for clarity).

Both the laser/television and thermal imaging sensor windows incorporate heaters to prevent fogging during system use. The thermal imaging sensor window has a heater that surrounds the window periphery. The laser and television sensor window has a very thin metal coating across the entire window surface. The metal coating is thin enough to be transparent, yet thick enough to conduct electrical energy and heat the window through resistance heating.

After additional analysis and confirming experiments, the SLAP problem-solving team concluded that the two windows' heaters had significantly different optical effects on their respective windows. The laser/television window heated the window evenly (due to the uniform metal coating used as a resistive heater). The thermal imaging sensor's germanium window heater, however, did not heat its window evenly. Due to the thermal imaging sensor window's peripheral heating approach, the window experienced uneven heating (the outer edge of the window was heated more than the center). The effect was that during heating the thermal imaging sensor window distorted, shifting the thermal imaging sensor optical path away from the laser and television optical path.

The problem-solving team did not immediately understand why this situation had not surfaced earlier. Upon further investigation, the team discovered that the window heater control software had been modified after completion of the SLAP development program. The software modification (which was directed by the customer) changed the temperature at which the window heaters triggered.

Prior to the customer-directed change, the window heaters energized at a temperature of 71° F (or lower), and maintained the windows at 71° F. After the software that controlled the window heaters was changed, the window heaters turned on at 105° F (or lower), and maintained the windows at this higher temperature. Note that the SLAP window heating feature was not evaluated during normal system acceptance testing at the General Digital manufacturing facility in California. Acceptance testing does not usually evaluate all design characteristics. Normally, the purpose of acceptance testing is to assure the product was correctly built.

The Team's Conclusions

After spending several months and significant funding to analyze the designation pointing error system failure, the team identified the window heater control software as the cause of the problem. Corrective actions included modifying the window heater software such that the window heaters held the window temperature at 71° F (as had been the case in the original configuration), which eliminated the designation pointing errors.

Problem Solving

One of the things that makes high technology quality management and continuous improvement simultaneously stimulating and frustrating is what, at times, seems like a never-ending problem stream. Strong problem-solving skills are essential to successful quality management implementation, and without these skills, one is doomed to solving the same problems repeatedly (or at least attempting to do so). Problem solving sounds like it might be a formidable task, especially in high technology situations as described above for the General Digital Suspended Laser Acquisition Pod.

Complex systems and processes are often encountered in modern industry, and the problems created by these systems and processes can seem intimidating, but the simple truth is that a basic problem-solving approach can be used to solve all problems. The technologies can differ (you might work in a business that requires expertise in sheet metal fabrication or financial analysis instead of optical physics), but the approach to solving problems for continuous improvement remains the same.

The Problem-Solving Process

A company president in a medium technology, highly successful company once shared his philosophy for solving problems with us.

He explained that he first learned of this approach as an undergraduate at the California Institute of Technology, and that the process had served him well in all aspects of his life (indeed, this man took his company from $12 million in annual sales to nearly $400 million in annual sales in about ten years, and this company produced a very high quality product). The problem-solving process our associate suggested consists of four steps:

1. What is the problem?
2. What is the cause of the problem?
3. What are the potential solutions?
4. What is the best solution?

The concept is shown in Figure 5-3, and explained in detail below. Note that the first two steps address root cause identification, and the last two steps address the development and implementation of effective corrective action. A root cause can be defined as the underlying condition that creates the problem. A corrective action is a solution to the problem (corrective actions are the actions that eliminate or control the root cause). With these definitions in mind, let's walk through the four step problem solving process.

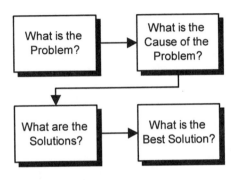

Figure 5-3. The 4-Step Problem Solving Process. Defining the problem is critically important and is all too frequently performed poorly, with predictably adverse effects on identification and selection of proper corrective actions.

Step 1: What is the problem?

On the surface, this appears to be an easy question to answer. It seems obvious this question should be answered before beginning the search for solutions.

Unfortunately, this doesn't always occur. More often than not, many of us attempt to define solutions before we really understand the problem. Consider the SLAP situation described above. John Lavery and the technical crew at General Digital spent a great deal of time and energy attempting to correct the SLAP designation pointing error. They thought that all of the systems sensors did not track with the line of sight for several months before they realized that the problem could be more accurately defined as the thermal imaging sensor not tracking with the line of sight. This might appear to be a subtle difference, but its ramifications were significant. The problem-solving effort stretched out considerably because the team focused on the wrong problem (one that was too broadly defined).

Here's a suggestion: The next time you are with a group of people working on a problem, observe the group interactions to determine how much time is spent on defining the problem accurately before attempting to identify and select solutions. If your experiences mirror ours (and we suspect they will), you might be surprised at how little time we tend to focus on that critical first step: defining the problem.

Step 2: What is the cause of the problem?

Once the problem has been satisfactorily defined (and remember, this is not a trivial thing to do), one can then move on to identifying potential causes of the problem. This is also a significant part of the problem-solving process, as we too often tend to immediately converge on a single cause without considering all potential causes. In the next chapter, on systems failure analysis, several sophisticated analytical techniques for identifying all potential causes will be developed. In most cases, though, a simple brainstorming approach will work. For now, it's only important that we recognize that it is not a good idea to jump to conclusions about a cause we intuitively feel is the underlying reason for a problem, but that we instead work to develop all potential causes.

Once these potential causes have been identified, each can be evaluated for its relationship to the problem. This usually results in ruling most of the potential causes out, and converging on the most likely cause. Sometimes there may be more than one cause. Multiple causes may be necessary to induce the problem, or perhaps there are several independent causes, each of which could result in the problem. Knowing that these causes are present is important, because if only one cause is addressed (and the others remain), eliminating only the one cause will not eliminate the problem.

Let's return to the SLAP designation pointing error. What happened there? The team working to solve the SLAP problem did not zero in on the real problem until many weeks had elapsed, and because the problem had not been accurately defined the proposed causes were off the mark. The team spent much time investigating the SLAP boresighting system, the power supply, and the system's microprocessor. The team did not identify all of the potential causes (partly because the problem had not been adequately defined), but instead jumped immediately to those causes they thought likely while others were not developed.

Step 3: What are the potential solutions?

Once the cause of the problem has been identified, one can begin to intelligently develop potential solutions. Almost all problems have more than one potential solution. At this point, it is best to be an option thinker, and to develop as many solutions as possible. Sometimes this may consist of developing multiple potential solutions to address a single cause. In other cases, the problem may be the result of more than a single cause, and multiple potential solutions should be developed for each cause. We recommend developing as many potential solutions as possible for each problem cause, and then listing the advantages and disadvantages of each.

Step 4: What is the best solution?

Having identified the problem, the cause(s) of the problem, and all potential solutions (along with their advantages and disadvantages), the final step in the problem-solving process is to select the best solution. Defining the advantages and disadvantages of each potential solution will greatly assist this process. There are other factors to be considered in selecting the best solution, and these factors comprise what we call an order of precedence for selecting corrective action.

Corrective Action Order of Precedence

As mentioned above, there are usually multiple options for corrective actions. These options generally fall into distinct categories, and these categories range from most preferable to least preferable when selecting corrective actions. The most preferable corrective actions are those that eliminate the problem's root causes. In so doing, these corrective actions totally eliminate reliance on customers, assembly technicians, or other personnel to perform in a specific manner such that the problem is eliminated. Although the expression is potentially derisive, this is commonly known as making a product, process, or service "idiot proof." The least preferable corrective actions are those that do not eliminate the problem at its root cause, but

instead rely on people to perform special actions to guard against the problem recurring. The corrective action order of precedence (starting with the most preferable to the least preferable category of problem solutions) is described below.

- *Design Upgrades to Eliminate or Mitigate the Problem.* This category of corrective actions modifies the product, process, or service to eliminate the features that induced the problem. For example, in the SLAP system described earlier, the heater design was modified to eliminate germanium window optical warping and its attendant thermal imaging sensor optical path distortion. This solution eliminated reliance on people to correct the problem. When a product, process, or service is modified to eliminate the feature inducing the problem, the problem is attacked at its root cause.

- *Requirements Relaxation.* In many instances, a problem is only a problem because the product, process, or service does not meet a specification. For example, the problem may be that a process has a yield of only 94 percent, when a yield of 95 percent is required by the customer. One solution is to negotiate a relaxation of the requirement such that a yield of 94 percent becomes acceptable. Obviously, such an approach is not entirely responsive to the customer's requirements, and as such, it tends to depart from effective quality management philosophies that focus on customer needs and expectations. Nonetheless, under certain circumstances, requirements relaxation may be the only practical approach, and as such, where requirements relaxation makes sense it should be considered.

- *Training.* In many instances, problems can be eliminated by providing training to customers, assemblers, or other personnel to control the circumstances that could induce a problem. One company we worked with experienced multiple instances of hydraulic fluid leakage on an aerial refueling system during system checkout prior to delivering the product to the customer. Application of the four-step problem -solving process described in this chapter revealed that hydraulic assembly personnel were unfamiliar with proper hydraulic line installation procedures. After training the hydraulic assembly personnel on proper hydraulic fitting and line installation techniques, the leakage rate during system checkout dropped significantly. Training might appear to not be as desirable a solution as would be redesigning the system such that it becomes insensitive to installation technique, but redesigning the aerial refueling system in this situation was not considered feasible by either the company producing the system or the customer.

- *Additional Testing or Inspection*. Under certain circumstances, falling back on sorting good product from bad through additional testing or inspection may be the most expedient solution. Again, this is counter to the TQM prevention (rather than detection) philosophy, and for that reason, additional testing or inspection is usually not an acceptable way of doing business other than in unusual circumstances. To illustrate the concept, we recall an example in which a munitions manufacturer procured a large number of detonators and found during acceptance testing that some of them were defective. The defect was induced by concavity in the detonator output surface, which resulted in a failure to reliably initiate an explosion (in this case, that's what the device was supposed to do). The munitions manufacturer had to meet a tight delivery schedule with its customer, the U.S. Air Force. The solution to this problem was to inspect the entire group of detonators for surface flatness, and to only use those that were acceptably flat. That was the short term solution, and it was selected to allow continuing production until the problem was eliminated at its root cause. The long term solution (which eliminated the root cause) resulted from the munitions manufacturer working with the detonator manufacturer to modify the detonator production process. The process modification eliminated the process features that induced detonator concavity.

- *Cautions or Warnings*. Another category of solutions includes incorporating cautions or warnings on products or in related documentation to prevent problems. This is not a preferred solution, but there may be

circumstances that require such a solution. We've probably all seen high voltage warnings on electrical equipment. Wouldn't it be better to simply insulate the high voltage areas such that it became impossible to contact a high voltage surface? The answer, obviously, is yes, but doing so may not always be practical. Maintenance technicians may require access to the system when it is energized, and it may not be possible or practical to isolate or insulate high voltage surfaces inside an electrical device. Again, if it's practical to eliminate the problem through a design change, then that is the most preferable solution.

- *Special Operational or Process Actions*. The last category, and the least preferable from a long term perspective, is to rely on special operational or process steps as a problem solution. When the U.S. Army procured Beretta 9mm service handguns, the initial shipments experienced structural failures after firing only a few thousand rounds. The Army's interim solution was a special operational action, which required replacing the handguns' slides after firing a specified number of rounds. The Army's long term solution was to eliminate the metallurgical deficiency that allowed the failures to occur.

The important concept to consider in implementing the above corrective action order of precedence is that even though it may sometimes be necessary to implement less preferable solutions (due to economic or other reasons), in all cases the long term solution should migrate toward a product or process change that eliminates the root cause of the problem.

Implementing and Evaluating Corrective Action

Once corrective actions have been identified, evaluated, and selected, the final steps consist of implementing the corrective action and evaluating its effectiveness. This may seem so obvious as to almost be insulting, but our experience has shown that in many instances the last two crucial actions are left hanging. This is particularly true in larger organizations, where it's easier to make assumptions concerning other people implementing corrective action. We recommend the following:

- After the corrective has been selected, the team working the problem needs specific actions and responsibilities. Those so designated need to agree with the required actions and commit to implementing them.

- Once the corrective action is reported to have been implemented, the problem solvers need to follow up to assure that the corrective action is in place.

- Once the corrective action is confirmed as having been implemented, the problem solvers need to monitor the situation to assure that the problem has in fact been eliminated.

The above steps are critical, and we believe the last one is especially significant. The quality measurement system discussed in Chapter 4 helps tremendously in evaluating if recurrences are present after the corrective action has been implemented. Why focus on this? Can't we simply assume that the corrective action will correct the problem?

Our experience indicates the answer is a strong *no*. This can occur for a variety of reasons. Perhaps the team working the problem failed to adequately define the problem, and they solved the wrong problem (recall our admonition at the beginning of this chapter, in which we pointed out that people unskilled in problem-solving techniques frequently fail to adequately define the problem). Perhaps the four-step problem-solving process results in the group properly defining the problem, but they converge on the wrong cause. Perhaps something even more subtle occurred, and the team found a legitimate cause of failure, but they did not find all of the causes of failure. If corrective actions are not implemented to address all of the failure causes, the causes that are not addressed will continue to create failures.

Summary

This chapter developed a simple four-step problem-

solving process that begins with defining the problem, moves on to identifying the problem's causes, defines all possible solutions, and then selects the best solution. We developed a corrective action order of precedence, with the most preferable solutions being those that modify the process or product design such that the root causes of the problem are eliminated. This approach eliminates reliance on people assembling or using the product or service in a special manner to prevent problems from occurring. Where such design changes cannot be implemented (perhaps due to cost or other reasons), other solutions consisting of special steps, caution or warning labels, or additional inspection or testing may be more expedient. One should recognize, however, that these solutions do not eliminate root causes, and a more permanent solution should be pursued as a long term fix.

Finally, this chapter discussed the importance of seeing that problem solutions are implemented, and that once implemented, the problem is eliminated or at least its frequency of occurrence is satisfactorily reduced.

The next chapter builds upon the material presented in this chapter and develops more sophisticated failure analysis techniques when problems exist in complex systems or processes. Although the analysis techniques are more sophisticated, they will follow the same four-step problem-solving process outlined here.

References

Managing Effectively, Joseph and Susan Berk, Sterling Publishing, 1991.

"Problem Solving/Process Improvement," *Managing For Six-Sigma Quality*, General Dynamics Valley Systems Division TQM Training Materials, 1990.

Systems Failure Analysis, Joseph Berk, Managing Effectively Seminars, 1999.

Chapter 6

Systems Failure Analysis

Finding root causes in complex systems failures...

Kevin Arnot sat in front of Dr. Gary Sims' walnut desk. Dr. Sims, Allied Techsystems' president, sat behind the desk, lost in thought. Allied Techsystems is a small, high technology company that develops and manufactures ground-based air defense missile systems.

Dr. Arnot had been in many meetings with Dr. Sims, and knew not to interrupt him when the man's body language clearly indicated he was thinking. Arnot's boss, Dr. George Redding (the vice president of engineering), sat next to him. "Kevin," Dr. Sims finally said, "we've got a real problem here. You probably heard that during the 38th Brigades' annual service practice in Korea, one of our Avenger missiles detonated in flight prior to reaching the target drone. This is the first one we've lost in a long time, and I'm not sure what to do. We don't have any hardware to examine, because the missile exploded over the Yellow Sea. We're not even sure what the bird was doing when it went down. All the Army told us was that it suddenly went up. To hear them tell it, it almost sounds as if it received a self-destruct command, but we don't know that for sure."

"We have essentially no information other than that," Dr. Redding said. "It's going to be next to impossible to find out what went wrong here."

"Can we at least talk to the crew?" Arnot asked.

"Not for several days," Dr. Sims answered. "The Army is bringing them back to Fort Bliss, in El Paso, but we won't be able to talk to them for at least a week. The major concern now is the Army's fear that this condition could affect all of the Avenger missiles in Korea."

Arnot thought about the lost Avenger, and the lack of hardware to examine for the failure analysis. If the Avenger missile was not available, that would greatly impede any analysis. At least the team should be able to get the battery control van and the radars, Arnot thought. The battery control van was the ground-based enclosure from which the crew operated the missile system, and the radars sent signals from the battery control van to the inflight missiles.

The analysis would be complex, though, Arnot realized. The system included the missiles, with their own flight control and propulsion systems, radar energy communications with the battery control van, the control van itself, and all the interconnect cabling. The failure that caused the missile to detonate in mid-flight could have occurred in the missile, in the control van, in the radars, or in any of the other system elements that linked the Avenger and its ground systems. It could have even been a crew error. Perhaps someone in the command van accidentally depressed the self-destruct button. Arnot knew there were numerous possibilities, perhaps even hundreds of potential failure causes.

"How about the video in the control van?" Arnot asked.

"Same story," Sims said. "At least a week before it's back in the country, and another two days for the Army to censor it."

"Any other information?" Arnot asked.

"We have other Avenger missiles here, as I'm sure you already know," Sims said. "The Army wants missiles shipped to Korea next week, so you ... a few days to check out things on what we ...ve here before we ship them out."

Arnot looked at Redding and Sims. Both offered no clues as to what they were thinking, although it was clear both were concerned.

"Any other questions?" Dr. Sims asked Arnot.

"Not right now," Arnot answered, "but I probably will have some soon." Arnot left Dr. Sims' office and immediately began to plan the Avenger failure analysis. He knew that he would need to review the company's quality data base and field reports on all previous failures. He also knew that he needed to get quality assurance, manufacturing engineering, and procurement involved. About 80 percent of the Avenger's major components were procured from Allied Techsystems' subcontractors.

Arnot considered inviting an assembly technician from the production line, but he didn't know whether the problem originated in the missile, the battery control van, the radars, or elsewhere. Consequently, Arnot didn't know which area to invite a technician from, and he didn't want to have too many people in the meeting. Arnot thought about who might have a good feel for the systems interactions on a "hands on" basis, but he couldn't think of anyone.

At 1:00 p.m., the attendees arrived in Arnot's office. Arnot had a large plain paper pad and extra copies of the agenda. Arnot began the meeting immediately. "Our problem," he explained to the group, "is that an Avenger detonated in mid-flight, and we can't recover the hardware for the failure analysis. My thought is that we should prepare a fault tree analysis to guide our investigation, and that's why I wanted to get each of you here. I thought we might start out by reviewing what we know about the failure, and then we can start the fault tree."

"Sounds fine to me," the quality assurance manager said, and with that, Arnot began explaining what he knew about the failure from the morning's meeting with Sims and Redding.

In the last chapter, we outlined a four-step approach

for effective problem solving. In many organizations that manufacture or manage complex systems, a more sophisticated approach is required to arrive at the root causes of systems failures. Although the approach we are about to present in this chapter offers more sophisticated techniques for developing potential failure causes and required corrective actions, the technologies involved still follow the four-step problem-solving process described in the last chapter: defining the problem, defining all potential causes of the problem, developing potential solutions, and selecting the best solution. Figure 6-1 outlines the systems failure analysis approach, and its relationship with the four-step problem-solving process.

Systems Failure Analysis

- Begins with Clear Understanding of Failure Symptoms
- Objectively Identifies All Possible Causes
- Objectively Evaluates Likelihood of Each
- Converges on Most Likely Cause(s)
- Identifies All Potential Corrective Actions
- Evaluates Each Potential Corrective Action
- Selects Optimal Corrective Action(s)
- Evaluates Efficacy of Corrective Action(s)

The Four-Step Problem Solving Process

Figure 6-1. The Systems Failure Analysis Approach. The steps involved in a complex systems failure analysis are based on the four-step problem-solving process presented in Chapter 5.

Before delving too deeply into this subject, a few definitions are in order. A systems failure occurs when a system does not meet its requirements. A laser failing to designate its target, an aerial refueling system failing to transfer fuel at the proper flow rate, an integrated blood chemistry analyzer failing to provide accurate test results, a missile that detonates prematurely, and other similar conditions are all systems failures.

A systems failure analysis is a study to determine the underlying reasons for the nonconformance to system requirements. A systems failure analysis is performed for the purpose of clearly identifying the root causes of the nonconforming condition and recommending appropriate corrective actions.

Root causes are the underlying physical conditions

or procedures that result in the nonconformance. This concept of a failure root cause is important, and it has to be distinguished from a mere symptom of the failure. The root cause is what induces the failure; a symptom is one of the conditions that results. Failure to understand this distinction is really nothing more than a breakdown in the four-step problem-solving process. Those who don't make the distinction often won't be able to solve the problem (or perform an accurate failure analysis) because they are working the wrong problem.

Our last definition addresses corrective action, and it is the same as that presented in Chapter 5. Corrective action is a change to a system's design, its manufacturing processes, or its operating or maintenance procedures to eliminate or mitigate the effects of the failure's root cause.

Figure 6-2 shows a flow diagram for performing systems failure analysis. Systems failure analysis begins with a clear understanding of the failure symptoms (which is really nothing more than defining the problem). The process then objectively evaluates each of the potential failure causes. In a few pages, we'll discuss the fault tree analysis technique, which is an invaluable tool for identifying potential failure causes. Once all potential failure causes have been evaluated, the systems failure analysis process objectively evaluates the likelihood of each potential cause.

We'll use a tool called the failure mode assessment and assignment matrix (or FMA&A, for short) along with several supporting analytical techniques to accomplish this. Having identified and evaluated all potential failure causes, the process allows converging on the most likely causes. Frequently, more than one likely cause emerges, and we'll review techniques for handling this situation.

Once the most likely causes are known, the systems failure analysis process then identifies all potential corrective actions, evaluates the desirability of each corrective action, and selects the optimal corrective action. As discussed in Chapter 5, there is an order of precedence in selecting corrective actions. Those that modify the system design or the process such that the root causes are eliminated are most desirable; those that rely on special actions or procedures are least desirable (although we should

recognize that in many cases, interim fixes consisting of special actions or procedures, or other "Band-Aid" fixes, may be necessary). Finally, the process ends when post-corrective-action implementation shows that the system failure no longer recurs.

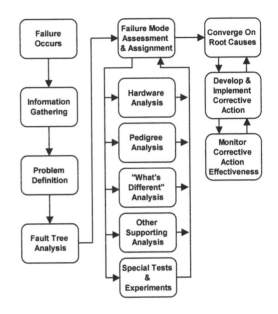

Figure 6-2. The Systems Failure Analysis Process. The fault tree analysis develops all potential failure causes, the failure mode assessment and assignment matrix lists each of the potential failure causes and actions required to evaluate them, supporting analysis techniques allow converging on the most likely failure cause, and corrective actions are implemented and evaluated for their effectiveness in eliminating recurrences.

Identifying All Potential Failure Causes

When confronted with a systems failure, we've observed a natural tendency to begin disassembling hardware to search for the cause. This is a poor approach. The Japanese believe that a failed piece of hardware is the most valuable hardware available, as it can reveal vital secrets about what caused the failure. We concur with that assessment.

Recognizing that failed hardware can reveal valuable information, one must also recognize that safeguards are necessary to prevent losing that information. If one immediately attempts to disassemble a failed system without knowing what to look for, the risk of missing vital clues or destroying valuable evidence is high. For these reasons, it is necessary to know what one is looking for prior to disassembling failed hardware. How can we determine what to look for? This is where the fault tree analysis enters the picture.

Fault tree analysis is a graphical technique that develops all potential causes for a failure. The approach was developed in the early 1960s by Bell Laboratories, working with the U.S. Air Force and Boeing, on the Minuteman missile development program. The Minuteman is an intercontinental nuclear missile. When the Air Force and Boeing were developing this weapon system, they were quite concerned about inadvertently launching a nuclear missile. The Air Force and Boeing wanted a technique that could analyze the missile, its launch system, the crew, and all other aspects of the complete weapon system to identify all potential causes of an inadvertent launch. Bell Laboratories developed the fault tree for this purpose.

The fault tree starts with a top undesired event, which is simply the system failure mode for which one is attempting to identify all potential causes. From a systems failure analysis perspective, the fault tree's top undesired event is the same thing as the first step in the four-step problem-solving process. It's the definition of the problem. The analysis then continues to sequentially develop all potential causes. We'll examine a simple example to see how this is done in a moment, but first, let's consider the fault tree analysis symbology.

Figure 6-3 shows the symbols used by the fault tree. There are two categories of symbols: events and gates. Let's first consider the four different symbols for events. The rectangle is called a command event, and it represents a condition that is induced by the events immediately below it (we'll see how shortly). The circle represents a basic failure event (these are typically component failures, such as a resistor failing open, or a structural member cracking). The house represents a normally occurring event (for example, if electrical power is normally present on a power line, the house would be used to represent this event). The last event symbol is the diamond (it looks like a rectangle with the corners removed), which can represent either a human error or an undeveloped event. A human error might be a pilot's failure to extend the landing

gear when landing an aircraft, a technician's failure to properly adjust a variable resistor, or a crew member inadvertently depressing a self-destruct button on a missile control console. An undeveloped event is one that requires no further development. Usually command events considered extremely unlikely are designated as undeveloped events to show that they have been considered and eliminated as possible failure causes.

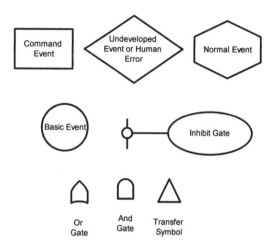

Figure 6-3. Fault Tree Analysis Symbology. The symbols represent different events that could cause the failure being analyzed, as well as the relationships between these events.

Fault tree events are linked by gates to show the relationships between the events. There are two types of gates: "and" gates, and "or" gates. The "and" gate signifies that all events beneath it must occur simultaneously to result in the event above it. The "or" gate means that if any of the events beneath it occurs, the event above it will result. The concept is further clarified in Figure 6-4.

Who should participate in developing the fault tree? The best approach is to assemble a team (as will be described in subsequent chapters) consisting of personnel with a good understanding of how the system is supposed to operate. This would typically include an engineer, a quality engineer, a manufacturing engineer, an assembly technician, and perhaps others, depending on the nature of the failure.

Let's now examine how all of the above comes together to generate a fault tree analysis. We'll consider a simple system failure analysis. Suppose we have a system with a light bulb that screws into a socket, and the light bulb illuminates when someone turns a switch on. Figure 6-5 shows a schematic for this system. One day, you flip the switch, expecting the light to work, but the light does not come on.

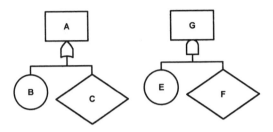

Figure 6-4. Fault Tree Gates. The "and" gate requires all events below it to occur simultaneously to result in the event above it. As shown above, Events E and F must occur for Event G to result. The "or" gate means that the event above it will result if any of the conditions below it are present. As shown above, if either Event A or Event B occurs, Event C will result.

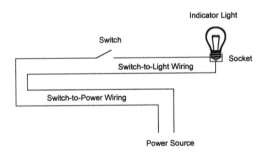

Figure 6-5. Indicator Light System. In this system failure, the light bulb failed to illuminate. A fault tree analysis is developed to identify all potential causes of this failure.

The first step in solving any problem is to define the problem. What is the problem here? The light bulb does not illuminate. This also becomes the top undesired event in the fault tree for this system failure, and Figure 6-6 shows it in a command event (the rectangle symbol). Top undesired events are always shown in a command event symbol, as they will be commanded to occur by the events shown below it.

The next step in performing a fault tree analysis is to look for the immediately adjacent causes that can induce the event shown in the rectangle. This is a critically important concept. A common

shortcoming is to jump around in the system, and start listing things like a power loss in the building, a failed switch, and perhaps other events, but the fault tree requires discipline. One has to look for the internal or immediately adjacent causes. An approach for doing this is to imagine yourself as the light bulb, screwed into the socket, and ask "What can happen *in me or right next to me* to prevent me from illuminating?"

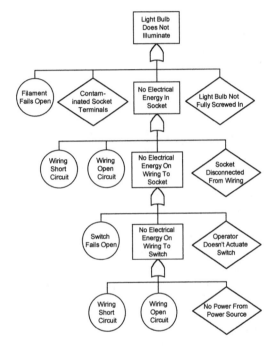

Figure 6-6. Indicator Light Fault Tree Analysis. This fault tree develops potential causes for the light bulb shown in Figure 6-5 failing to illuminate.

If one considers only these conditions, the answers are:

- An open light bulb filament.

- Contaminated terminals in the socket.

- A bulb that's not fully screwed into the socket.

- No electrical energy from the socket.

The next step in constructing the fault tree is to show these events immediately below the top undesired event, and then to determine which

49

symbol is appropriate for each. The open filament is a basic component failure, so it goes in a circle symbol. Contaminated terminals in the socket could be caused by a variety of conditions, but for the purposes of this analysis we won't fully develop these, and we'll put contaminated terminals in an undeveloped event symbol (the diamond). Not fully screwing the bulb into the socket is a human error, so it goes into a human error symbol (also a diamond). Finally, no energy from the socket is a condition that will be commanded to occur if other events occur elsewhere in the system. This event becomes a command event, and it goes into a rectangle. All of these are shown in Figure 6-6.

The above events are all of the internal or immediately adjacent conditions that can cause the light bulb to fail to illuminate, and this nearly completes the first tier of the fault tree for this undesired event. To complete this tier, we have to link these internal and immediately adjacent events to the command event above them. Either an "and" gate or an "or" gate will be used. The question here is: Will any of the events below the top undesired event result in the top undesired event, or are all events below the top undesired event required to result in the top undesired event? In this analysis, any of the events below the top undesired event will result in the light bulb failing to illuminate, so the "or" gate is selected.

The fault tree analysis continues by developing the potential causes for the next command event, which in this case is the event that appears below the top undesired event on the fault tree's first tier: No electrical energy available in the socket. We now need to identify all conditions internal to and immediately adjacent to the socket. In this case, the socket can be disconnected from the wiring, or the wiring can have no power delivered to it, or the wiring could have a short circuit, or the wiring could break open. The socket being disconnected from the wiring would probably be the result of a human error, so this event is shown in a diamond. The wiring having a short circuit is a basic component failure (the wiring insulation fails), so it is shown in a circle. The same is true for the wiring failing open. No power to the wiring is commanded to occur by conditions elsewhere in the system, so it is shown as a command event. Any of these conditions can cause the command event

immediately above it (no electrical energy in the socket), so an "or" gate is used to show the relationship.

We continue the analysis by identifying all internal or immediately adjacent conditions to the wiring. The internal wiring conditions have already been addressed (the wiring failing open or having a short circuit), so only the immediately adjacent conditions need be shown. This brings us to the next element in the circuit, which is the switch.

The switch can fail open (this is a basic switch component failure, so it's shown in a circle). The switch can have no power delivered to it (this is a command event).

Finally, the system operator might forget to turn the switch on (this is a human error). Any of these conditions can induce the "no power to wiring" command event shown immediately above these events, so an "or" gate is used.

The same kind of wiring failures are shown for the wiring leading from the electrical power source to the switch (similar to the types of failures we developed for the wiring earlier). The only other condition that can cause no power to this wiring is no power from the source. Since there is no information about the power source, it is shown as an undeveloped event. Any of the conditions can induce no power on the wiring, so an "or" gate is used.

At this point, the fault tree logic for our simple example is completed. What does this mean? With the data available, the fault tree started with a definition of the failure (which became the top undesired event in the fault tree) and systematically developed all potential causes of the failure. It's important to note that the fault tree logic development started at the point the failure appeared (in this case, a light bulb that failed to illuminate), and then progressed through the system in a disciplined manner. The fault tree logic followed the system design. Systematically working from one point to the next when constructing a fault tree forces the analyst to consider each part of the system and all system interfaces (such as where the switch interfaced with a human being). This is key to successful fault tree construction, and in taking advantage of the fault tree's ability to help identify all potential failure causes. Before leaving the fault tree, there's one more task, and that's assigning a unique number to each of the basic events, human errors, normal events, and undeveloped events. These will be used for tracking purposes, as will be explained in the next section.

The FMA&A

After completing the fault tree, the next step is to initiate preparation of the failure mode assessment and assignment matrix (or FMA&A). As Figure 6-7 shows, the FMA&A is a four column matrix that identifies the fault tree event number, the fault tree event description, an assessment of the likelihood of each event, and what needs to be done to evaluate each event. The FMA&A becomes a table based on the outputs of the fault tree that lists each potential failure mode.

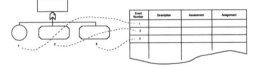

Failure Mode Assessment & Assignment Matrices
What To Do With FTA Findings

The Approach:

- **Number FTA Basic, Human Error, Undeveloped and Normal Events**
- **Carry Events to FMA&A Matrix**
- **Use Matrix To**
  - **Make Assignments for Evaluating Each Event**
  - **Track Event Evaluation Progress**
  - **Record Event Evaluations**
  - **Guide Failure Analysis Meetings**
  - **Converge on Failure Cause(s)**

Figure 6-7. Fault Tree Analysis and Failure Mode Assessment and Assignment Matrix Relationship. The FMA&A is keyed to the fault tree analysis. It defines the actions necessary to evaluate each hypothesized failure cause identified by the fault tree.

The third column in the FMA&A, the likelihood assessment, lists the failure analysis team's opinion on each potential failure cause being the actual cause of the failure. Usually, failure analysis teams list each hypothesized failure cause as likely, unlikely, or unknown. When the FMA&A matrix is first prepared, most of the entries in this column should be listed as unknown, since at this point no work beyond the fault tree construction and initiation of the FMA&A has been started.

The last FMA&A column (the assignment column) lists the assignments agreed to by the failure analysis team members to evaluate whether each hypothesized failure mode actually caused the observed failure. This ties back to our earlier discussion in which we advised against tearing a system apart immediately after a failure without knowing what to look for (i.e., what might have caused the failures). The fault tree and FMA&A provide this information. These analysis tools reveal to the analysts what to look for when disassembling the system, as well as when conducting other activities to evaluate the likelihood of each potential failure cause.

The assignment column of the FMA&A defines the actions necessary to look for and determine if each hypothesized failure cause did or did not contribute to the failure. We recommend that failure analysis team members also indicate in the fourth FMA&A column the team member who has responsibility for each assignment and required completion dates for these assignments. The assignment column is also used for general comments describing the team's progress in evaluating each event, significant findings, and other progress-related information. A review of this column should provide a general indication of an ongoing failure analysis' status.

Most people find it effective to update the FMA&A during a failure analysis meeting instead of taking notes, editing the FMA&A after the meeting, and distributing the FMA&A well after the meeting has concluded. Most word processing packages include a tables feature, and if possible, a computer should be used during the meeting to update the failure analysis status in real time. In this manner, an updated FMA&A can be printed immediately at the end of each failure analysis meeting, which helps to keep the failure analysis team members focused and sustain the failure analysis momentum.

Figure 6-8 shows an FMA&A prepared for the indicator light fault tree analysis developed in the preceding pages. The FMA&A shown in Figure 6-8 shows what actions are required for evaluating each indicator light potential failure cause, and it provides a means of keeping track of the status of these actions.

Event Number	Description	Assessment	Assignment
1	Filament Fails Open	Unknown	Examine bulb for open filament. Rodriguez; 16 March 93
2	Contaminated Socket Terminals	Unknown	Examine socket for contaminants. Perform FTIR analysis on any contaminants observed in socket. Rodriguez; 16 March 1993
3	Light Bulb Not Fully Screwed In	Unknown	Inspect bulb in socket to determine if properly installed. Smith; 14 March 1993.
4	Socket Disconnected from Wiring	Unknown	Examine wiring and perform continuity test. Ashoggi; 16 March 1993.
5	Wiring Short Circuit	Unknown	Examine wiring and perform continuity test. Ashoggi; 16 March 1993.
6	Wiring Open Circuit	Unknown	Examine wiring and perform continuity test. Ashoggi; 16 March 1993.
7	Operator Does Not Activate Switch	Unknown	Interview operator and check switch function. Rodriguez; 16 March 1993.
8	Switch Fails Open	Unknown	Check switch function. Rodriguez; 16 March 1993.
9	Wiring Short Circuit	Unknown	Examine wiring and perform continuity test. Ashoggi; 16 March 1993.
10	Wiring Open Circuit	Unknown	Examine wiring and perform continuity test. Ashoggi; 16 March 1993.
11	No Power from Power Source	Unknown	Check power supply with multimeter. Ashoggi; 14 March 1993.

Figure 6-8. Indicator Light Failure Mode Assessment and Assignment Matrix. This FMA&A matrix shows the actions necessary to evaluate each potential indicator light failure cause identified by the fault tree.

Organizing the Failure Analysis

This chapter has introduced several concepts thus far, with the fault tree analysis and FMA&A figuring as prominent tools for identifying potential failure causes and organizing the effort to evaluate each. The importance of a multidisciplinary team participating in the failure analysis was covered here and in Chapter 5, and the team concept will be further developed in Chapter 8. Prior to explaining the supporting analyses used for evaluating each of the hypothesized failure causes, let's examine how a failure analysis effort should be organized.

Figure 6-9 shows a concept we find works well. Once a failure occurs and an assignment to investigate it is made, all related data should be gathered and the team members identified. We recommend convening the team members, familiarizing the group with the failure and what

information is available, and then preparing the fault tree. We also recommend preparing at least a preliminary version of the FMA&A in this first meeting, and negotiating assignments and required completion dates prior to adjourning the meeting. If the failure analysis is for a failure that has interrupted production or has severely impacted a customer, we recommend that the failure analysis team reconvene on a daily basis, updating the FMA&A with the results of the failure analysis team members' assignments. This will allow the group to begin to converge on the most likely cause (or causes) of failure.

Supporting Analyses

Once the failure analysis team has completed the above steps, the team has a fault tree that identifies each potential failure cause, and an FMA&A that provides a road map and management tool for evaluating each potential cause. At this point, it becomes necessary to turn to the family of supporting technologies shown in Figure 6-2 to complete the FMA&A, and in so doing, converge on the true cause of the failure (or causes; sometimes more than one cause may be present).

As Figure 6-2 shows, there are a number of supporting technologies to help evaluate each of the potential failure causes.

These include the "what's different?" analysis, the pedigree analysis, special tests, and finally, analysis of the failed hardware. Each of these is explained below.

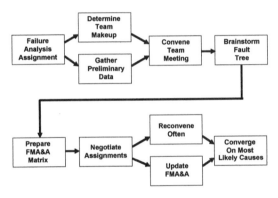

Figure 6-9. Organizing the Failure Analysis. This approach helps to bring the right information and team members together to prepare the fault tree and FMA&A, and to pursue a systematic identification and evaluation of all potential failure causes.

"What's Different?" Analysis

The "what's different?" analysis is a technique to identify any changes that might have induced the failure. The basic premise of this analysis is that the system has been performing satisfactorily until the failure occurred; therefore, something must have changed to induce the failure. Potential changes include such things as system design, manufacturing processes, suppliers, operators, hardware lots, or perhaps other factors.

The "what's different?" analysis will almost certainly identify changes. As changes are identified, these should be evaluated against the potential failure causes identified in the fault tree

and the FMA&A. One has to do this in a systematic manner. Keep in mind that changes are always being introduced, and when a change is discovered, it doesn't necessarily mean that it caused the failure. Design changes can be identified by talking to the engineers assigned to the system. Procurement specialists should be asked to talk to suppliers, as changes may have occurred in procured components or subassemblies. We've found that it also makes sense to talk to the people responsible for maintaining the engineering drawings (this function is normally called configuration management or document control). The people responsible for maintaining the engineering design package normally keep records of all design changes, and they can frequently identify changes the design engineers may not remember.

Manufacturing process changes are sometimes more difficult to identify. Assembly technicians, their supervisors, and inspectors may be able to provide information on process changes. The manufacturing engineers and quality engineers assigned to the system may have information on process changes. Most companies have written work instructions, and sometimes records are kept when these documents are changed, so the work instruction history should also be researched. Tooling changes can also be identified in this manner. Companies often keep tooling release records, so these, too, should be reviewed. Here's another tip: Look around the work area for tools not identified in the assembly instructions or otherwise authorized. Sometimes manufacturing personnel use their own tools, and these might be changed without any documentation.

If the system in which the failure occurred was manufactured in a facility that uses statistical process control, process changes may be more readily available. Ordinarily, process changes are noted on statistical process control charts. This subject is further developed in a later chapter.

Another area to search for potential changes is in the system operating environment. Sometimes a subtle environment change is enough to induce a failure. Recall the Suspended Laser Acquisition Pod failure discussed in Chapter 5. That failure's root cause was a design change related to the window heating temperature, but it took a change in the operating environment to make the failure emerge.

As an aside, it almost makes sense to evaluate the environment in which the failed system attempted to operate against the environmental capabilities of the system and all of the components that could have contributed to the failure (as identified by the fault tree). Even though the system may have been operating in its specified environment when the failure occurred, it is not unusual to discover that as a result of a design oversight, one or more of the system's components or subassemblies is being asked to operate outside of its rated environment.

Before leaving the "what's different?" analysis, a few caveats are in order. Recall that at the beginning of this discussion we identified a basic premise, which was that a change occurred to induce the failure. This may not be the case. Sometimes nothing has changed and the failure resulted from normal statistical variation (this concept is covered in more detail in a later chapter). There's another possibility, and that is that the failure may have been occurring all along, but it was not previously detected.

Pedigree Analysis

Pedigree analysis examines all paperwork related to the components and subassemblies identified in the fault tree and the FMA&A. Due to normal quality assurance and other record keeping requirements, most companies have a fairly extensive paper trail on the hardware used in modern manufacturing operations. Pedigree analysis involves studying this paperwork to determine if it shows that the components and subassemblies identified in the fault tree and the FMA&A met their requirements. This paperwork can include test data, inspection data, raw material data sheets (from the raw material suppliers), and other certifications.

The data described above should be examined for any evidence that shows nonconformances or other inconsistencies. Although it may seem incredible, it's not uncommon for test or inspection sheets to show that a part or subassembly did not meet its requirements (i.e., the item was mistakenly accepted).

Again, one has to be on guard for unrelated findings. We recommend only examining the data for the parts or subassemblies identified by the fault tree

analysis (and consequently, the FMA&A). Since the fault tree identifies all potential causes of the observed failure, it isn't necessary to explore other data. Here's another caution: If a nonconformance is indicated in the data sheets, it should be addressed, but the nonconformance has to be compared to the fault-tree-identified potential failure cause. Again, just because a nonconforming condition is found, it doesn't mean it caused the failure. We've found that when performing pedigree analysis, one often finds other problems. These need to be fixed, but they may not be related to the systems failure being analyzed.

Pedigree analysis should also review prior quality records (for the system and all components and subassemblies identified by the fault tree and FMA&A as potential failure causes) to identify any prior similar failures. In many cases, specific systems failure modes will not be new, and reviewing the findings of previous analyses will support the failure analysis process.

Special Tests

Special tests can include tests designed to induce a failure or other designed experiments. Tests designed to induce a failure can often be used to evaluate a hypothesized failure cause. By configuring the hardware with the hypothesized failure cause present, one can determine if the previously observed failure mode results (i.e., the one that initiated the systems failure analysis). The disadvantage of this type of test is that it tends to require absolute failure causes (in other words, the hypothesized failure cause must induce the observed failure mode all of the time). In many cases, hypothesized failure modes can induce the observed failure mode, but they may not do so all of the time. Another disadvantage of this type of test is that it can typically only evaluate one hypothesized failure cause at a time. In so doing, the effects of other contributory conditions often cannot be evaluated.

Taguchi tests are useful for evaluating the effects of several variables and their interactions simultaneously with a minimum number of test samples (we'll cover this topic in a subsequent chapter). Frequently, when evaluating all of the hypothesized failure causes in the FMA&A, one can eliminate most of the causes, but several will remain. In many instances, more than one of the remaining hypothesized failure causes contributes to the failure. Because the Taguchi test technique evaluates several variables simultaneously, it can be used for assessing the relative impact of each of the remaining potential failure causes.

Hardware Analysis

Throughout this chapter, we have emphasized the importance of deferring analysis of the failed hardware until all of the potential failure modes have been identified, and the failure analyst knows exactly what to look for in the failed hardware.

The fault tree analysis and FMA&A allow development of a logical hardware teardown and inspection process. We recommend using this information to prepare written disassembly instructions and an inspection data sheet. Photodocumenting the disassembly process and all hardware as the analysis progresses also makes sense. The analyst may later become aware of additional potential failure causes, and photodocumenting the teardown process allows one to re-examine the hardware as it was removed from the system.

There are several tools available to assist the hardware analysis. These fall into several categories: magnification equipment, material analysis equipment, general measuring equipment used for evaluating compliance to engineering design requirements, and other specialty equipment.

Optical microscopes and scanning electron microscopes are two magnification tools that have gained wide acceptance. Optical microscopes are available in most companies, and they permit greatly magnified observations of suspect components. Sometimes tiny defects or witness marks that cannot be seen with the naked eye are visible under magnification (these defects or other marks often support or refute hypothesized failure causes). If greater magnification is required, the scanning electron microscope has proven to be a valuable tool. Scanning electron microscopes bounce electrons off the surface being examined to produce an image, and they can magnify images up to one million times their actual size. Scanning electron microscopes are generally available in larger

companies, but if your organization does not have this capability, there are many commercial metallurgical and other failure analysis laboratories that do.

The most common material analysis tools are energy dispersive x-ray analysis, Fourier transform infrared spectroscopy, and other forms of spectroscopy. A detailed description of the operating principles of these tools and their limitations is beyond the scope of this book. The point to be made here is that tools are available (perhaps not in your organization, but certainly in others specializing in this business) that can help you to evaluate if the correct materials were used, or if contaminants are present.

General measuring equipment consists of the standard rules, micrometers, gages, hardness testers, scales, optical comparators, and coordinate measuring machines available in almost every company's quality assurance organization. This is where quality assurance failure analysis team members can lend tremendous support. For most systems failures, the fault tree analysis and FMA&A will identify many components, which, if nonconforming, could have caused the failure. The quality assurance organization can support the failure analysis by inspecting the components hypothesized as potential failure causes to determine compliance with the engineering design.

X-ray analysis is frequently useful for determining if subsurface or other hidden defects exist. X-rays can be used for identifying weld defects, internal structural flaws, or the relationship of components inside closed structures. This technique is particularly valuable for observing the relationships among components in closed structures prior to beginning the disassembly process. Sometimes the disassembly process disturbs these relationships (especially if the structure was not designed to be disassembled), and x-rays can reveal information not otherwise available.

Videotaping or filming moving machinery is another specialized technique that can reveal interactions not readily apparent when the machinery is stationary. Many videotapes and film projectors allow the film to be shown in slow motion, which also helps to evaluate potential failure modes. Computer technologies and digital imaging are combining to provide very rapid movie capabilities for desktop computer analysis, which can provide even more advanced capabilities.

All of the failed hardware should be identified and bonded so that it is available at a later time if the need arises. Sometimes new potential failure modes are identified as the analysis progresses, and it might become necessary to re-evaluate hardware previously examined. We recommend keeping failed hardware until corrective action has been implemented and its effectiveness in eliminating failures is confirmed.

Returning to Allied Techsystems

Recall the situation outlined at the beginning of this chapter, and think about Dr. Sims' failure analysis challenge on the Avenger missile. He doesn't have any failed hardware to evaluate. How can he progress? Two points are significant. The first is that failed hardware often does not reveal the causes of failure, so not having the actual failed hardware available for review is not necessarily a handicap (in fact, this is frequently the case for aircraft, missile, and ordnance failures). The fault tree analysis and FMA&A have will reveal what could have caused the failure. Frequently, other supporting analyses (the "what's different?" analysis, the pedigree analysis, and others) will reveal conditions that caused the failure without the failure analysis team having to examine the failed hardware.

The second point is that even if the actual failed hardware is not available, other hardware may be. Recall from the conversation between Dr. Sims and others that other Avenger missile system hardware was available for examination. The cause of failure for the missile that detonated in flight might not be revealed in this other hardware, but the other hardware offers opportunities for related hardware analysis. Many of the missile's and the other system elements' subsystems and other components might similarly be available for examination (perhaps the organization has these in house as field returns or new inventory).

Moving Toward Failure Analysis Completion

As the potential failure causes identified by the fault tree analysis and the FMA&A are evaluated, what typically occurs is most of the potential failure causes are ruled out. In most cases, a few potential causes remain. One approach is to perform additional specialized testing, as discussed earlier, to converge on the actual cause or causes of failure. Another approach is to implement a set of corrective actions that addresses each of the remaining unconfirmed potential failure causes. Either approach is acceptable, although one should take precautions to assure that the selected corrective actions do not induce other system problems. Corrective action selection, implementation, and evaluation should be based on the findings of the failure analysis.

In most cases, failure analysis teams will find a confirmed failure cause during their systematic evaluation of each potential failure cause in the FMA&A. A word of caution is in order here, though. The natural tendency is to conclude the failure analysis as soon as a confirmed failure cause is found without continuing to evaluate the remaining potential failure causes contained in the FMA&A. We strongly advise against concluding the failure analysis until all potential causes are evaluated. The reason for this is that multiple failure causes frequently exist. In a recent circuit card failure analysis, the failure analysis team performed a fault tree analysis and identified 87 potential failure causes. In systematically evaluating each of these, the team found that six of the causes were present. Any one of these failures was sufficient to induce the failure mode exhibited by the circuit card. If the team stopped when they confirmed the first cause, the other five causes would remain, and the circuit card failures would continue to recur.

Creating and Using Failure Analysis Reports

Recording systems failure analysis results is important for several reasons. A well-written failure analysis report details the results of the failure analysis, and it defines required corrective actions. The failure analysis report provides a permanent record of the failure, the analysis to identify its causes, and required corrective actions. This information can greatly accelerate future failure analysis efforts.

We recommend a narrative format for a systems failure analysis, organized as described below:

- *Executive Summary*. This section should provide a brief description of the failure, its causes, and recommended corrective actions. We recommend including it on the first page of the failure analysis report.

- *Description of the Failure*. This section should provide a detailed description of the failure, and the circumstances under which it occurred.

- *Conclusions and Recommendations*. This section should present the failure analysis findings and recommended corrective actions. The information should be more detailed than that provided in the executive summary.

- *"What's Different?" Analysis*. This section should define any differences discovered during the analysis that are related to the failure. If there were no differences related to the failure, the report should explain the differences that were found, and why they were considered to be unrelated.

- *Pedigree Analysis*. This section should define the documentation reviewed during the failure analysis, and the conclusions of the review.

- *Environment Analysis*. This section should define the environment in which the system was attempting to operate when it failed. It should also define the rated environments of all of the system's components and subassemblies (i.e., those identified in the fault tree analysis and FMA&A as potential contributors to the failure). This part of the analysis should state whether the system's operating environment contributed to the failure.

- *Hardware Analysis*. This section should define the results of the hardware failure analysis. We recommend including photographs or other illustrations supporting the findings of the failure analysis.

- *Prior Failure History*. This section should describe any previous similar failures (if any exist), and prior corrective actions. The relationship between the prior failures and the one being analyzed should be described.

- *Appendices*. We recommend including the fault tree analysis and the FMA&A as appendices to the failure analysis report.

We also recommend maintaining a library of prior failure analysis reports as part of the quality measurement system data (as discussed in Chapter 4). As mentioned earlier, these analyses will help future failure analysis efforts. A library of failure analysis reports can also be used to proactively support an organization's continuous improvement efforts in other areas. New development efforts should incorporate the wisdom gained from prior failure analyses to assure inclusion of design and process features that will prevent similar failures. The failure analysis library should also be reviewed for applicability to an organization's other products to prevent similar failures from occurring on these.

Summary

In many companies that manufacture complex systems, a more sophisticated approach is required to arrive at the root causes of systems failures. The approach presented in this chapter offers sophisticated techniques for identifying potential failure causes and evaluating corrective actions, but the technologies involved still follow the four-step problem-solving process: defining the problem, defining all potential causes of the problem, developing potential solutions, and selecting the best solution. Fault tree analysis is used to define all potential causes of failure, and the FMA&A is used to guide the systematic evaluation of each of these potential causes. Several supporting technologies are used to assist in this systematic evaluation. The systems failure analysis process outlined in this chapter allows converging on the most likely failure causes. Frequently, more than one likely cause emerges, and several techniques were described for handling this situation. Once the most likely causes are known, the systems failure analysis process then identifies all potential corrective actions, evaluates the desirability of each corrective action, and selects the optimal corrective action in accordance with the guidance of Chapter 5.

References

Managing Effectively, Joseph and Susan Berk, Sterling Publishing Company, 1997.

"Ordnance System Failure Analysis," Joseph Berk and Larry Stewart, *Proceedings of the Annual Reliability and Maintainability Symposium*, 1987.

Fault Tree Construction Guide, Armament Development Test Center United States Air Force, May 1974.

Fault Tree Analysis, Waldemar F. Larsen, Ammunition Reliability Division, Picatinny Arsenal, United States Army, August 1968.

Components Quality/Reliability Handbook, Intel Corporation, 1985.

Chapter 7

Employee Involvement and Empowerment

People really know the problems and how to solve them...

Al Seidenstein was frustrated. He watched as the stamped sheet metal computer case fell out of its forming die, knowing that the stamping was cracked and would have to be scrapped. Seidenstein's frustration was only partly due to his knowledge that this particular case would be scrapped. Seidenstein's real frustration grew out of the fact that nearly 30 percent of the cases he stamped ended up in the scrap bin. Seidenstein knew how to fix the problem, but no one would listen to him.

Seidenstein knew the corners of the case were too sharp. The sheet steel either had to be heat treated to make it less brittle prior to stamping, or the design had to be relaxed to allow a more generous corner radius. Seidenstein had worked in the forming area for more than 15 years, and he knew that when sheet steel stampings started cracking near bends, it had to be due to the material being too hard, or the bends being too sharp, or a combination of both factors.

Seidenstein had relayed this information to his supervisor more than a year ago, but nothing had been done to fix the problem. Seidenstein remembered the conversation with the supervisor well. The supervisor listened intently to Seidenstein's recommendation. He then explained that the last time an employee offered a suggestion for fixing a quality problem, management shut the area down for six weeks to implement the improvement. They fixed the problem, but the shop floor workers were off the payroll for six weeks, and when they came back, two positions had been permanently eliminated. No one in the work center really saw the benefit in that upgrade.

Seidenstein understood the shop floor supervisor, but he reasoned that with proper planning, the problem could probably be corrected in less than a couple of days. Seidenstein submitted a suggestion through the employee suggestion program, but he never heard anything about his recommendation (not even an acknowledgment that it had been received).

Seidenstein picked up the stamping as it fell from the die and examined the edges where the sheet metal formed the case's corners. More cracks. He made a note on his scrap report and tossed the stamping into the scrap bin. Seidenstein thought about what people in the metal recycling business earned. He wondered if it paid better than his job as he loaded another sheet of steel into the stamping machine.

The People Side of the Equation

One of the greatest underlying factors in the success or failure of any organization is the power of its people, and how well that power is focused toward meeting the organization's objectives. As our society moves in the direction of automated processes, computer-controlled machinery, and computer-driven manufacturing resource planning, most of us tend to think that our reliance on people has decreased. Science fiction has often conjured images of totally automated factories devoid of people, churning out products in a highly structured environment. Modern manufacturing management pursues the goal of a paperless factory, with design concepts moving from an engineering computer-aided-design terminal through data links to a computer-aided-manufacturing terminal, which in turn drives a numerically controlled machine.

Some of these visions are closer to reality than many

of us might realize. We recently visited a very efficient electronics manufacturing plant operated by Martin Marietta in Orlando, Florida. The Martin Marietta plant operates with very few people. Robots carry supplies from stockrooms along designated paths to work centers. Workers rely on computer terminals for work instructions. Automated test equipment tests the product and reports the test results.

The above automation example notwithstanding, all companies operate on the strengths and weaknesses of their employees. Even in the Martin Marietta plant described above, employees have to design, maintain, and operate the systems that create output. Organizations that can tap the strengths of their people will be stronger and more competitive than those that cannot. Organizations that regard people as automatons or mere cogs in a wheel will never realize their full potential. In the long run, such companies' inefficiencies attract competition, and unless the management philosophy changes, they will disappear.

Involvement and Empowerment

A fundamental TQM precept is that employees must be involved, and they must be empowered. What do these terms mean?

Employee involvement means that every employee is regarded as a unique human being (not just a cog in a machine), and that they are involved in helping the organization meet its goals. It means that each employee's input is solicited and valued by his or her management. It means that employees and management recognize that each employee is involved in running the business.

Employee empowerment is a somewhat different concept. It means that in addition to involving employees in running the business, employees and management recognize that many problems or obstacles to achieving organizational goals can be identified and solved by employees. Employee empowerment means that management recognizes this ability, and provides employees with the tools and authority required to continuously improve their performance. It means that management states its expectations about employees recognizing and solving problems, and empowers them to do so.

Facilitating Employee Involvement

We believe that most managers want to have their employees involved in improving the business, or at least to be an active participant in helping the business meet its objectives. In many organizations, however, this is not true for all employees. In every organization it's possible to identify people who make things happen, and others who are along for the ride. It's possible to identify people who are well suited for the work they are doing (and who enjoy their work), and others who seem to enjoy their work less, and perhaps are not so well suited for it.

An earlier chapter in this book discussed the 80/20 concept, and this concept can be applied to the employees described above. We've all been in organizations where 20 percent of the people in an organization do 80 percent of the work. Imagine what would happen in those organizations, though, if everyone became as enthusiastic and productive as the 20 percent that do 80 percent of the work. That 20 percent would maintain their output, but now the other 80 percent of the people would increase their productivity. The mathematics of the situation indicate an organization's output might increase by a factor of four or more. Employee empowerment is a management philosophy that allows increases of this magnitude to emerge.

Facilitating employee involvement requires recognizing the value of each individual, understanding human motivations, assigning people to positions in which they can be successful, and listening to employees. Let's explore each of these topics, and how they help to assure employee involvement.

Round Pegs in Round Holes

Let's consider another all-too-common scenario (and one that we have probably all observed). Have you ever seen a situation in which a marginally performing employee is reassigned to a new job and suddenly becomes a superstar? We recall an instance in which a procurement specialist for stamped metal parts was regarded as a marginal performer (the kind large companies only keep around because it's easier to wait until the next layoff than take action now). This employee (we'll call him Doug) was assigned to a value improvement team (of the type to be discussed in Chapter 14), and he suddenly came alive. Doug's ideas on reducing costs through improved procurement strategies emerged daily. Doug started visiting suppliers and showing them how to increase their productivity (thereby lowering their cost and price while increasing their profit). Doug became a key member of a team that increased his employer's profitability, when only a few weeks earlier he had been slated for a layoff.

What happened in the above example? Doug came alive because he was involved and empowered. He was placed in a position for which he had both aptitude and interest, his boss was interested in what he was doing, and his boss gave him the green light to fix problems. His boss didn't tell him what to do, he only told him what needed to be accomplished and empowered him to make it happen. Doug became a example of the productivity gains that result when people are assigned work they want to do, and are given the freedom to do it.

What Motivates People?

One of the most important tasks faced by any management team is motivating their organization's members. Understanding what it takes to motivate people is an important element of empowerment. Without motivation, all the empowerment in the world will do no good, so it is necessary to first understand what motivates people.

Understanding motivation is not a simple thing to do. We all have different motivations. The situation is complicated by the fact that there are no simple answers in defining what motivates people. There are several models to describe human motivation. The two that come closest to modeling the human behaviors (in our opinions) are Maslow's hierarchy of needs and McClelland's more contemporary model for describing motivation.

Maslow developed his hierarchy of needs model in the 1940s. The Maslow model is based on an ascending order of human motivations, starting with the most basic motivation of survival, and then progressing upward through safety, love, esteem, and self-actualization (Figure 7-1 shows the concept).

Figure 7-1. Maslow's Hierarchy of Needs. The Maslow model of human motivation is based on an ascending order of needs, starting with survival and progressing up to self-actualization.

Maslow's model stipulates that as each need is satisfied, one progresses up through the hierarchy of needs. In other words, one faced with starvation or exposure would only be motivated by attaining those goals that would eliminate the threat of physiological destruction. A person in this situation would not be too concerned about higher order motivations, such as love, or esteem. Once survival becomes a relatively sure thing, though, one moves up the hierarchy of needs and take steps to assure one's own safety. Once that need has been met, one continues to move up the hierarchy of needs to search for love, and then the esteem of one's associates, and then finally, self-actualization (or, to borrow a phrase, to be all that one can be).

The Maslow hierarchy of needs model also works in reverse, but it skips steps. One who has satisfied higher levels immediately reverts to focusing on a lower-order motivation if it becomes threatened. A man who writes books for self-actualization forgets his self-actualization needs and immediately reverts to a safety or survival motivation if a thief robs him at gunpoint.

McClelland's later work (in the 1960s) built a model to describe the motivations of people who had already satisfied their needs for safety and survival. McClelland recognized three motivations: achievement, affiliation, and power. McClelland's model is similar to Maslow's in that it recognizes that the security factors have to be satisfied first (the need for survival and safety). The McClelland model is significantly different than Maslow's after these basic needs are met, however, in that it postulates that the needs for achievement, affiliation, and power are not hierarchical, but are instead present in all people in differing amounts. The McClelland model holds that we all have needs for achievement, affiliation, and power (Figure 7-2 shows the McClelland model).

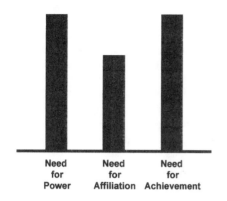

Figure 7-2. The McClelland Motivation Model. McClelland theorizes that needs for power, affiliation, and achievement are present in all of us, but in varying degrees.

What are these three needs of achievement, affiliation, and power? The need for achievement is self-explanatory. It's the motivation that inspires each of us to create, to attain objectives, to meet sales quotas, or to satisfy the other personal conditions that represent achievements to each of us. The need for affiliation is the need to be liked by others, or to associate with others. It's the motivation that causes us to look forward to an organization's social interactions, to join associations, or to take other steps that allow us to be with our fellow human beings. The need for power is not, as its name implies, a need to direct the efforts of others or to dominate their lives. It is instead perhaps more appropriately described as a need for influence, or for us to modify our surroundings such that they are more acceptable. This need for power (or influence) is often high in those leading an organization, and they certainly influence their surroundings. It's often also very strong in their advisors (those who most strongly influence the boss).

What does all of the above mean to us? What it means is that we empower people for a purpose, and

this purpose is to allow them to do their jobs in a more efficient, higher quality, and hopefully, more enjoyable manner. As stated earlier, all of the empowerment in the world will be of little use if we don't understand what motivates individuals. The emphasis has to be on the word "individual," because each of us is unique, and each of us has different combinations of motivations. To implement TQM through employee involvement and empowerment, one has to recognize these individual motivations, and create an environment that allows individuals to satisfy these motivations.

Listening: A Basic Involvement Tool

A colleague of ours for whom we have enormous respect once commented that most people have two ears and only one mouth, and (in his words) "there's probably a message there." The clear implication is that most of us might be well served by spending at least twice as much time listening as we do talking.

Based on our observations, listening is one of the most effective tools managers can use to promote employee involvement. This may sound trivial, but it is not. Listening to a human being is a powerful involvement tool. It helps the speaker feel that the person listening wants to understand what the speaker has to say. It encourages people to open up, and to become involved. If no one listens, people won't become involved, because they will recognize (properly so, in our opinion) that no one values their opinions.

There are a few techniques for effective listening that are really no more than common sense and good manners. The first, and perhaps most obvious, is to listen. When someone is speaking, listeners should refrain from speaking. Imagine what would happen if a manager asked an employee for an opinion on something, and as soon as the employee started to speak, the manager jumped in and explained the "real" problem, and what needed to be done to fix it. When this occurs, in one act management is simultaneously telling the employee that his or her opinion is valueless, management has all the answers, and no one really cares what the employee thinks, anyway. Unfortunately, most of us don't have to think very hard to recall examples of such behavior. The great tragedy here is that the people we should be listening to often understand the real problems, and know what needs to be done to fix them.

The next step in good listening is to ask questions, but to do so in a nonthreatening and open-ended manner (open-ended questions are those that require more than a yes or no answer). Suppose a work center is producing an unacceptably high number of nonconforming parts. One approach to soliciting employee input in such a situation is to ask the employees "What kinds of tools do you need to make your job easier?" This is a good approach, as it does not hammer the employees for producing nonconforming material, it clearly conveys management commitment to support the work center with whatever it needs to continuously improve, and it induces the employees to start speaking up about needed improvements.

Another approach (and one that should be avoided) is to ask "What are you doing wrong? Why are you making so many nonconforming parts?" This sort of negative questioning is threatening, and few of us are eager to share what we are doing "wrong" with management.

It's also a good idea to ask questions from time to time when employees are explaining their ideas for improving an operation to assure that you understand what they are saying. This tends to help the speaker understand that you really are interested in what he or she has to say.

Another topic has to do with taking notes. There are two schools of thought here. Note-taking can be an intimidating thing (it tends to make the dialogue more "official," and for that reason, it may inhibit the speaker). In other cases, it further reinforces your commitment to fully understanding what the employee has to say. Our recommendation is to play it by ear. If employees seem inhibited if you start to take notes, then don't. If they seem encouraged, then take notes to heighten your understanding.

One last thought on listening has to do with summarizing what you think you have heard. This will help to make sure that you have heard what the employee has to say, and gives the employee an opportunity to correct any misperceptions on your part (and to provide more information). We've also found that summarizing in this manner at the end of a conversation further reinforces to the employee that management is committed to hearing what the employee has to say.

Suggestion Programs: Another Involvement Tool

Many organizations use formal suggestion programs with varying degrees of success as a tool to facilitate listening to employee suggestions. The format for these programs usually involves suggestion boxes and forms throughout the facility, with periodic management review of the suggestions and feedback to the people making the suggestions. Many companies have also incentivized the process, offering cash or other awards for approved cost reduction or quality improvement suggestions.

A suggestion program, however, is no substitute for listening directly to employees about their ideas for improvement. It's one thing to put up a few suggestion boxes and hope for input. It's quite another to actively and directly solicit input from employees.

Life After Listening: What Comes Next?

We believe that the most important element of listening to employees (be it through formal suggestion programs or simply meeting with employees to hear their suggestions) is following up on every suggestion and improvement recommendation. Not every suggestion will be implemented, but all should be answered. The simple act of listening to employees raises their expectations for improvement. Listening and then failing to provide feedback on the status of a suggestion or improvement idea is probably worse than not listening at all. If managers don't listen, employees will only suspect that management doesn't value their ideas. If managers listen and then fail to provide any feedback, the suspicion will be confirmed.

Sometimes managers are afraid to provide a negative response to an employee recommendation. We've observed that most employees are not offended by a rejection of their suggestions if the idea is not feasible, the reasons for the rejection are explained, and the explanation is offered in a constructive and appreciative manner.

Facilitating Employee Empowerment

The preceding pages addressed employee involvement. Let's now turn to the other topic in this chapter, which is employee empowerment. Facilitating employee involvement requires all of the topics discussed above, plus a willingness for management to cede some of their authority to their subordinates. Employee empowerment requires developing a set of continuous improvement objectives that employees will accept. Employee involvement also requires that management provide employees with the tools necessary to do their jobs, and to continuously improve their jobs.

Ceding Authority and Overcoming Fright

Empowering employees to fix problems is a frightening concept to many managers, especially those in first-level supervision (i.e., those directly supervising people who do the work). The fear is that in giving employees the authority to make decisions and solve problems, managers might be relinquishing their authority, and perhaps, their value to the organization. Overcoming these fears is critically important, because if first-level supervisors or other managers are opposed to employee empowerment, it will not work. If this fear exists in your organization, it exists for a reason, and that reason is usually because senior management has the same fears about middle management, and middle management has the same fears about first-level supervision. When the problem emerges at the first-level supervisor-to-worker interface, though, it's usually more recognizable because the workers are usually the ones we want to empower.

When this kind of resistance is encountered, we've observed that the usual reaction is for senior and middle management to blame first level supervisors for opposing the program. If the organizational culture is such that employee empowerment is generally accepted and the fear exists only in individual managers or supervisors, it's more appropriate to work the problem on an individual basis.

In either case (either organizational or individual resistance to employee empowerment), the approach to overcoming this rather significant barrier is the same. Our suggestion is to emphasize that the organization's value system rewards those managers who develop their employees and who empower them, rather than simply giving direction and expecting it to be executed. Organizations with cultures that run counter to such a value system will find that the cultural shift is easier to describe than accomplish. Consistency in values and senior management commitment will assure success, but such a change can take years to implement. Patience is required.

One of the manifestations of management's reluctance to let go can be observed when managers embark on an employee empowerment effort, but then define both the problems to be solved and how to solve them. Telling people they have the authority to define problems and implement corrective actions makes no sense if management then proceeds to do both (with the "empowered" employees serving only as observers). Employees will see through this situation instantly. The result will be just the opposite as that intended. Employees will withdraw, instead of becoming involved and empowered.

A Common Vision: Defining Expectations

There's a saying we've mentioned elsewhere in this book: If you don't know where you're going, you'll probably end up somewhere. Management has an obligation to prevent that from happening in an organization, and that responsibility is particularly germane to the issue of employee empowerment and continuous improvement. Management must define where the organization is going and what it expects from the organization's members to prevent it from "ending up somewhere."

How does this relate to employee empowerment? Having authority with no sense of direction related to where the authority should be applied is useless. It makes no sense to empower employees if they do not know management's expectations. If employees do not understand the overall direction in which the organization is headed, they will not understand why they are empowered, or how or where to apply this empowerment. This also relates to the discussion on maximizing the output of all employees, as mentioned earlier in this chapter. Management has a responsibility to define a common vision for the organization so that empowered employees understand where the organization is going, and what their roles are in attaining this common vision.

Providing the Right Tools

Most of us have heard the Old Testament story about the Hebrew slaves in ancient Egypt being commanded to make bricks with no mortar or hay. Empowering employees without providing them with the right materials to do their jobs makes about as much sense (indeed, it would probably take divine intervention for employees in such a situation to be successful).

Management has a responsibility when it empowers employees to implement continuous improvements to provide employees with the tools necessary for doing so. Some of these tools involve developing the necessary skills through training (a subject to be covered shortly). Others are literally tools, such as new machinery to implement process improvements, new material handling equipment to eliminate material handling damage, new forms to facilitate increased efficiencies, or any of a variety of other tools identified by employees as necessary to improve their operations. Asking employees for improvement recommendations and then failing to provide the required tools is not empowerment at all (rather, it is an exercise in frustration for the "empowered" employees). As mentioned earlier in the discussion on listening to employees, it will not always be possible to implement every employee improvement concept, but when this is the case, management still has an obligation to respond to the request, and to explain the reasons it cannot be implemented.

Training

One of the essential tools for empowering employees is training, especially in the areas of identifying problems and implementing continuous improvement. Training related to these needs is generally available from consultant or training organizations. Some companies opt to offer the training internally, with internal training experts or other personnel tapped to prepare and present the training sessions. Regardless of the source, a successful employee involvement and empowerment effort requires providing problem-solving skills.

Rewarding Success

Here's another old saying: Nothing succeeds like success. This is particularly true in the employee involvement and empowerment business, and the message here is that when success is achieved, publicize and reward it. Companies attempting to implement or improve their employee involvement and empowerment efforts frequently find that if their internal success stories are publicized, others in the organization soon get the message that management is committed to giving employees the tools to fix problems, and that the effort is producing tangible results.

A Few Obstacles and Recommended Solutions

There are a few things that can quickly kill an employee involvement and empowerment effort. Some of them have already been mentioned in this chapter, and others are mentioned elsewhere in this book. We'll reiterate a few of the more significant risks to involvement and empowerment efforts here, both to reemphasize their importance and to provide recommendations on avoiding or overcoming them.

One of the most foolish things management can do is to adopt any hint of negativism in reviewing employee recommendations for improvement. If the idea does not make sense, explain why in honest terms. If employees are adamant about the improvement recommendation's soundness, reconsider it with an open mind. We've seen more than a few improvement recommendations implemented after initially being disapproved.

Management negativism can creep into a discussion in strange ways. Consider what happens when management is exposed to a very significant employee-recommended improvement, and then comments (in all innocence, usually with no malice) "I can't believe we didn't see this sooner." What happens to the employees who developed the improvement? They will probably feel inadequate for not having developed the idea sooner. What happens to the employee's managers and supervisors? They feel even more threatened. After all, they managed and supervised the now-recognized-to-be-inefficient area (before the continuous improvement recommendation came to light). Masao Nemoto, formerly a senior executive with Toyota Motors, points out that comments of this sort are very deleterious to a continuous improvement effort, and must be avoided.

Fear is another strong negative motivation, and it should also be avoided. Management experts all over the world are in agreement on this subject. Deming even lists it as one of his major quality points. Any employee involvement and empowerment effort that is attempted over a foundation of fear will collapse.

It's tough to find a quality management consultant these days who will admit to favoring slogans, but it's been a fact of American manufacturing management that quality slogans seem to barrage us endlessly. Some people even regard the phrase "total quality management" as a slogan. Even though quality slogans have been an integral part of quality improvement efforts for years, there is essentially no research to suggest that a slogan has ever made a difference in an organization's quality.

Deming advises against the use of slogans, and we do, too. We believe that slogans are insulting to the intelligence of those to whom they are targeted (typically, those who do the work). The people who do the work know what the problems are, and they usually have a pretty good idea about what's necessary to correct the problem (and the required corrective action is not a slogan). The bottom line: Stay away from slogans. They trivialize a challenge that is too significant to wish away with a chant.

Failure to respond to employee recommendations, as mentioned earlier in this chapter, is another sure-fire way to kill an employee involvement and empowerment effort. If management does not acknowledge employee recommendations, employees will rapidly conclude that management has no interest in their ideas. As stated earlier, management must acknowledge all improvement recommendations, including the ones that are not deemed feasible. The United States Army Air Defense School at Fort Bliss, Texas, has an excellent approach for acknowledging suggestions from the troops. All suggestions initially receive a letter acknowledging their submittal. The letter's heading is "A Penny for Your Thoughts," and the acknowledging letter has a brand new penny taped to it. Corny? Absolutely, but it works. Soldiers

know their ideas are heard and are being evaluated.

Another way to kill an employee involvement and empowerment effort is to punish anyone as a result of a continuous improvement recommendation. While this seems so unlikely a course of action as to hardly be worth mentioning, let's consider the earlier examples. Suppose a manager comments that a suggestion makes good sense, but expresses disappointment that the idea had not been recognized and implemented earlier. Comments of that nature are essentially reprimands, and should be avoided at all costs.

Returning to Mr. Seidenstein

Let's go full circle and return to Al Seidenstein, with the defective metal stamping dilemma described at the start of this chapter. Seidenstein's problem with defective metal stampings was not being addressed even though he knew how to fix the problem. One of the reasons his supervisor was not too anxious to move out on Seidenstein's ideas was the result from a previous suggestion: a six-week shutdown (without pay) while the suggestion was being implemented and the permanent loss of two jobs. Certainly, such actions are perceived negatively by employees and will put a damper on their willingness to become involved in developing concepts that might result in loss of income.

Continuous improvement will result in improved efficiencies, and improved efficiencies often mean the elimination of positions within an organization. Harley-Davidson, a premier motorcycle manufacturer, recognized this potential detractor early in their mid-1980s continuous improvement and profitability turnaround. Harley-Davidson solved the problem by working in concert with their labor union to effectively prevent the elimination of jobs mandated by efficiency improvements. Harley-Davidson's approach was to pull in work formerly subcontracted to other companies (thereby creating more jobs within Harley-Davidson) to keep those displaced by efficiency improvements on the payroll. The approach is working well.

The bottom line is that organizations seeking to involve and empower their employees in an ongoing quest for continuous improvement must recognize that positions may be eliminated as a result of the improved efficiencies (and employees will know this). The organization has to have an approach for allaying these fears.

Summary

This chapter reviewed two concepts critical to total quality management implementation success: employee involvement and empowerment. Employee involvement means tapping the strength of every employee and facilitating their participation in helping the organization meet its goals. Employee empowerment means that management recognizes the ability of employees to define problems and implement continuous improvements, and provides employees with the tools and authority required to do so.

References

Total Quality Control For Management, Masao Nemoto, Prentice-Hall, Incorporated, 1987.

Well Made In America: Lessons From Harley-Davidson On Being The Best, Peter C. Reid, McGraw-Hill Publishing Company, 1990.

Organizational Behavior and the Practice Of Management, David R. Hampton, Charles E. Summer, and Ross A. Weber, Scott Foresman, and Company, 1978.

The Deming Management Method, Mary Walton, Perigee Books, 1986.

"Suggestion Systems Gain New Lustre," Michael A. Verespej, *Industry Week*, November 16, 1992.

Chapter 8

Corrective Action Boards and Focus Teams

One plus one is always greater than two...

Ray Garcia examined the composite structure on the table in front of him with a mixture of frustration and disappointment. As the manufacturing engineer assigned to Maxwell Aviation Composite Components' F-22 structures program, he had been struggling with a delamination problem for several weeks. Maxwell Aviation was barely able to keep up with its production delivery schedule, but it was doing so at great expense. Almost a third of the structures being fabricated on the F-22 program showed evidence of delamination after curing and had to be scrapped. Maxwell Aviation's customers made it clear they would not accept structures that had been repaired. The structures either met their requirements without repairs or they went in the scrap heap.

The delamination problem was particularly frustrating for Garcia, as he considered himself to be an above average manufacturing engineer. In Garcia's three years of manufacturing engineering experience, he had not encountered a process problem he could not resolve. Garcia didn't know what to do next. He had tried changing the work instructions to add more resin. When thát failed he tried less resin. The delaminations continued. He next experimented with different cure times, specifying longer and then shorter cycles in the oven after the filament wound structure had been assembled. This had not worked, either. Garcia then suspected that the operators were not following the work instructions. He audited the process, and found the operators to be in total compliance with his documented work instructions. Garcia had even ordered new lots of resin, filament thread, and the other materials used for making the F-22 structures, but nothing had made a difference.

Garcia felt he had nowhere to go. He wanted to see his boss, Jim Trent (the manager of manufacturing engineering at Maxwell Aviation), but Garcia was afraid Trent's opinion of his engineering skills would be lowered. Garcia finally decided that he had run out of new things to try, and if he didn't see Trent soon, the boss would soon be looking for him about the low yield. Garcia went to see Trent that afternoon.

"Hello, Ray," Trent said as Garcia entered his office. "The guys tell me we're still having problems with the F-22 delaminations. Making any progress in fixing it?"

"Well, I've tried several approaches," Garcia answered. "I've altered the resin content, and I've tried changing the cure times, and..."

"Nothing changed, though?" Trent asked.

"No," Garcia said, looking at the floor.

"Hmmm. What do you think you ought to do next?" Trent asked.

"I guess I don't know," Garcia said, fully expecting to be chastised for admitting it.

"Yeah, that's happened to me a lot as a manufacturing engineer," Trent said. "It's a lot easier sometimes to design these things than it is to figure out how to make them."

Garcia looked up, surprised at what Trent had just said. "You've had problems like this before?" he asked.

"Only three or four hundred times, maybe," Trent answered.

"What do you think I should do?" Garcia asked. He was surprised that Trent, whom he considered to be the brightest manufacturing engineer he had ever met, had encountered unsolvable process problems. Trent had his full attention now. "Should I try a different resin?"

"Beats me," Trent answered. "I don't have any idea what's causing the delaminations. You're a lot more familiar with the problem than I am, so I'd guess you would have a better handle on what to try next, but you don't, and what that's telling both of us is you probably need to ask somebody else who knows more than either of us about this problem what to try next."

"Do you think we need to bring in a consultant? Maybe see if we can find an outside expert?" Garcia asked.

"Yeah, I do," Trent said, "but you don't need to bring one in. You already have several on the payroll waiting to help you."

"Who?" asked Garcia.

"Well, you've got Tom Firestone, the designer," Trent said, "and Bill Pashon, the quality engineer, and Reza Fouza in procurement, and the guys on the line who are building the things. One of those guys may have the answer, or maybe not. But if you get all of them together, and they start bouncing ideas off of each other, the answer stands a pretty good chance of popping out. Why don't you get a team of our internal specialists working on this instead of taking the whole world on your shoulders, and see what happens?"

"Even the guys on the line?" Garcia asked.

"Especially the guys on the line," Trent answered. "They know more about what it takes to put those structures together, and what it takes to make them work, than you or I ever will."

"Is this what you've done when you couldn't solve a process problem?" Garcia asked.

"It's what I do when I can't solve any kind of problem."

Teamwork Synergy: The Power of People

In the past, and as part of our industrial culture, product or process problems were typically attacked by one or two people perceived to be the most knowledgeable in the area of concern. Sometimes these people were the managers of a particular work center. Many times they were assigned from outside the work center (for example, an engineer might be assigned to solve a problem in the machine shop). Sometimes they were technical experts brought in from outside the organization.

The above approach usually results in inefficient problem solving. Either the problem cannot be solved, or it takes far longer to solve than it should, or the solution is not preferred by those who have to implement and live with it. A more enlightened management approach (and one that is centered on a philosophy dedicated to continuous improvement) recognizes that problems usually cut across a company's organizational lines, and those closest to the problem are often best qualified to recognize and fix it. Manufacturing problems may be caused by a poor design. The people in manufacturing may be the only ones who recognize the problem. The existing process may not be adequate to meet required quality levels. Poor service can result from inadequate support documentation. Other examples can be developed indefinitely.

The previous chapter developed the concepts of employee involvement and empowerment. Employee involvement and empowerment have underlying themes including a sense of ownership, customer responsiveness, and continuous improvement. Most significantly, the employee empowerment aspect emphasizes giving employees the ability to implement actions consistent with these themes.

This chapter further develops these concepts by providing a structure for attaining synergy in the areas of employee involvement and empowerment through teamwork. This chapter will describe philosophies for creating and focusing teams that can solve problems and help organizations to meet their continuous improvement objectives in an efficient manner. We will then continue with a discussion of the formal requirements for Corrective Action Boards in a MIL-Q-9858 environment (primarily in systems defense contractors selling to the government, or to prime contractors who sell to the government), and the comparable Management Review required by ISO 9000.

Teams are a most powerful tool for solving problems and meeting continuous improvement objectives. Our country is only beginning to take advantage of the synergies available through teamwork. The American Society for Quality Control estimates that about 20 percent of the companies in this country has implemented employee-managed teams. That number seems low, but it represents a significant improvement in the last ten years (a decade ago, only about 5 percent of U.S. companies were using TQM-oriented teams). The bad news is that only 20 percent of companies are taking advantage of the synergies that can be attained through the use of teams. The good news is that the trend toward team-based problem solving is increasing, and 80 percent of our business enterprises have yet to take advantage of this powerful technique.

Two teaming concepts will be developed in this chapter: quality circles and focus teams. Although structures for both quality circles and focus teams are suggested, our readers should recognize that the quality circle and focus team concepts are similar (one is an outgrowth of the other), and the structures included in this chapter should be adapted to your organization. The idea is to start thinking in terms of the synergies that are available through teams (instead of simply designating one or two individuals to work problems), and implement the team concept in a manner that works for your organization.

Quality Circles

Quality circles comprise the most widely publicized team approach to problem solving. The technique originated in Japan in the early 1960s under the tutelage of Dr. Kaoru Ishikawa. Ishikawa figures prominently in many other TQM disciplines, including the development of the Ishikawa diagram, or as they are more commonly known, fishbone charts. These will be covered in more detail in

Chapter 10 in our discussion on statistical process control.

Ishikawa recognized the significance of shifting the responsibility for problem identification and problem solving to those on the factory floor. He knew that such an approach offered several inherent advantages. People on the factory floor are closest to the problems that interfere with delivering a quality product and meeting production schedules.

Ishikawa recognized that having personnel outside the work center identify and solve problems entailed another serious detractor, and that is change resistance. Externally developed solutions require work centers to accept and implement changes that work center personnel did not develop. This approach naturally results in resistance to change (and in particular, resistance to implementing a solution that might solve the problem).

Ishikawa reasoned that by including the personnel closest to the problem in the problem identification and solving process, two benefits would simultaneously be realized: People closest to the problem would be part of the team developing the solution, and by participating in developing the solution, the people closest to the problem would be more receptive to required changes.

Ishikawa had other objectives for Japanese industry, and these included educating line workers and their foremen in the quality assurance technologies. He felt that the quality circle concept offered an excellent vehicle for pushing problem identification, solution development, and corrective action implementation to the shop floor, while simultaneously educating workers and their foremen in quality-driven problem-solving technologies. Ishikawa even developed a Japanese quality assurance magazine, QC For Foremen, in 1962 to further push the quality circle concept.

The quality circle concept took root and grew rapidly in Japan during the 1960s and 1970s. American industry took a while to recognize the potential, but began to implement the concept in earnest in the late 1970s and 1980s. The concept has become popular in the United States, although it has disadvantages that can be overcome through more appropriately structured teams.

Quality Circle Organization

Quality circles were originally organized along the lines of the factory work center. Ishikawa's concept was that the people in a work center (along with their supervisor) would constitute the quality circle. The quality circle, Ishikawa taught, should be led by the work center supervisor. In that sense, the quality circle team would be exactly the same as the work center team. Ishikawa emphasized the importance of all work center members voluntarily participating in the quality circle. This early textbook approach to quality circle structures became one of its primary disadvantages. If the work center is not working together well, or if the supervisor is not well received by the people in the work center, a quality circle consisting of the same team members is not likely to succeed.

Ishikawa also emphasized the importance of giving the quality circle members the tools to meet their problem identification and problem-solving challenges. He emphasized the importance of training in quality assurance technologies (and in particular, problem-solving training) prior to expecting results from the quality circle.

How do quality circles work? Quality circles typically meet on a regular basis (usually once a week) to identify work center problems and develop solutions to these problems. The idea is to devote a specified period of time in which the work center group is not involved in their normal work center activities. This is often done during the lunch hour (with lunch either provided by the company or the group members bringing their lunch). Sometimes the meetings are held after hours, although that can create problems related to overtime pay, carpooling, and other potential detractors. The meetings are generally loosely structured, and often begin with a group brainstorming session to identify ongoing quality, schedule, and other problems being experienced by the work center. The meeting then turns to solving the problems identified by the quality circle members.

The Quality Circle Charter

The concept of continuous improvement was developed in Chapter 2. As discussed in that chapter, continuous improvement involves identifying problem areas, prioritizing these problems either by frequency of occurrence or cost, and then eliminating the problems' root causes. The quality circle charter is to implement continuous improvement using this approach. Ishikawa also believed, as do many of the organizations using quality circles, that the technique also offers a vehicle for self-development within the quality circle group.

Figure 8-1. Quality Circle Advantages and Disadvantages. The principal advantages of quality circles include synergy and employee involvement. The major drawbacks are inadequate focus and on occasion, a lack of technical expertise. Quality circles are usually limited to work center members, who may not possess all of the skills necessary to solve some problems.

Quality Circle Risks

Although the quality circle concept has gained considerable momentum in the United States, many organizations are finding that quality circles are not the panacea most people believed they would be. Quality circles offer a number of advantages, but they also have certain limitations, as Figure 8-1 shows. The problems most frequently encountered include a tendency for the quality circle concept to fade away if not nurtured, inadequate training, nonparticipation, inadequate technical expertise for the problems taken on by the quality circle, and a lack of focus (or working the wrong problems).

Maintaining Quality Circle Momentum

Quality circles are often not self-sustaining. Without management attention, the concept tends to die. The message here is that it is necessary for an organization's management to stay involved with its quality circles. Many organizations handle this by conducting periodic reviews with management to show the problems quality circles are working on and progress made in resolving them.

Some companies implement a standardized reporting format that allows senior management to quickly understand the problems being tackled by each quality circle (we'll say more on this when we discuss the focus team concept). A standardized reporting format forces quality circles to emphasize problems and solutions in a succinct manner.

Management, as always, must recognize that its role in reviewing quality circle progress is to provide support and encouragement. Any criticism can quickly undo a quality circle's momentum. This is especially true when the quality circle attacks a problem so obvious that management may tend to question why it hasn't been addressed earlier, as discussed in the previous chapter.

Quality Circle Training

Training appropriate for the problems being attacked by the quality circle is also essential. Simply directing people to identify and fix problems without providing the right tools is a near-certain formula for failure. One of the most essential training elements involves teaching people how to use quality measurement and other continuous improvement data (as described in Chapter 4) to assist in recognizing problems. Why is this necessary? Quality deficiencies that have existed for years may not be recognized as undesirable if the work group is conditioned to living with the problem.

Another important skill set that most people need to be trained to develop involves problem solving and failure analysis. Something that seems as intuitive as the four-step problem-solving process (described in Chapter 5) may not be known to people who have not been trained in how to deploy it. Most intuitive skills (like the four step problem solving process) only seem intuitive after one has been trained in their use. Organizations should also recognize that training is not a single-shot affair. Ongoing training in quality measurement techniques, problem solving, and other skills unique to the work center will go a long way toward sustaining quality circle efforts and enthusiasm.

In many cases, quality circles have taken on problems beyond their capability to solve. When quality circles are comprised of work center members only, some problems may require special expertise not present in the group. Management has to be aware of the problems quality circle members are working on to preclude this from occurring (if the quality circle members do not recognize a problem they cannot solve it). The risk is that the team wastes valuable time and becomes demoralized attempting to solve such problems.

Quality Circle Participation: Keep It Small

Active participation by the quality circle team members is also critical to success. This is an area in which most American organizations have departed from Ishikawa's philosophy on complete work center quality circle involvement and participation. The problem of active participation is compounded when the work center is large (say, more than five or six people, as is frequently the case in American industry). Large groups tend to make better decisions, but the decision-making process tends to take longer as the number of participants increases, and many people find this frustrating.

A quality circle comprised of 15 or 20 members tends to stifle both initiative and individual participation in group discussions. Our observations (as well as the observations of others) show that in such settings most quality circle members tend to wait for the group supervisor to speak and provide direction on the problems to be solved and the approach to be pursued in solving these problems.

Our advice is to keep quality circles small (no more than five members). We also feel it is highly desirable to have individuals other than the group supervisor lead the quality circle teams. Doing so both stimulates creativity and provides for developing leadership skills in quality circle members. Adopting this approach requires some off-line reassurance for group supervisors, as many group supervisors may feel they are losing control of their work centers when others are chartered with leading problem-solving efforts. An approach we have found helpful is to explain to group supervisors that a significant element of their job involves developing their subordinates. Complimenting supervisors when a quality circle is successful, especially when the supervisor did not lead the effort, greatly supports this concept.

Even in small quality circles of the size described immediately above, one often observes problems in stimulating active participation by all quality circle members. These issues most frequently emerge when one of the more active group members complains about "doing all of the work." Our experience has convinced us that the best approach to correct this situation is to work off-line with quality circle team leaders to encourage them to identify the tasks they would like the nonparticipating quality circle members to take on, and then to provide coaching on how best to elicit this participation. We emphasize the use of action item lists, assignment matrices, and other management tracking tools. We also encourage quality circle team leaders to direct open-ended questions (i.e., questions that cannot be answered with simple yes or no answers, but rather, require more in depth responses) to the nonparticipating team members. The reader is invited to review *Managing Effectively* (particularly Chapter 5 on delegating and Chapter 14 on meetings) for additional insights on these topics.

Losing Focus: Working the Wrong Problems

Perhaps the biggest problem with quality circles (a problem that has unfortunately led to many companies abandoning the quality circle effort entirely) is the meandering and wasted motion that frequently results when the quality circle groups lack focus. Near the beginning of this section, we developed the concept of quality circle members identifying and selecting the problems to attack. If inadequate training accompanies the quality circle implementation effort, or if the problems selected by the team are too sophisticated for the quality circle members to take on (perhaps, for example, a very subtle design or process modification is needed), the quality circle's efforts will start to falter. Another manifestation of this shortfall occurs when the team selects a problem that is of no real consequence. Yet another occurs when the quality circle does not adequately plan how to resolve the problem.

Adequate management review can help to prevent the problems described above, but our experience and observations show that the quality circle effort is most effective when management works to provide adequate focus and technical support. To do this requires a logical approach to selecting the right problems to attack, and providing the appropriate technical resources. This leads directly into a discussion of focus teams, which is the next level of sophistication in team-oriented problem-solving.

A Nation of Specialists

Virtually every company in America is made up of groups of specialists: managers, human resources professionals, engineers, manufacturing engineers, quality engineers, inspectors, assembly technicians, solderers, machinists, press operators, painters, maintenance personnel, and others. The mix of personnel depends on an organization's product line, and the organizational area in which people work. In many cases, continuous improvement can only be realized by orchestrating a blend of these talents to bring the full force of an organization's skills to bear. One cannot ignore the special talents and insights of those closest to the problem (and they are usually the work center members), but these people need to augment their problem-solving activities with adequate technical support.

The Focus Team Concept

The highly publicized quality circle concept has enjoyed notable success, but in many instances quality circles have been only marginally successful due to their inherent disadvantages. Many companies that have experimented with quality circle programs have quietly dropped them. The inherent shortfall of the quality circle approach is that it typically lacks focus on the right problems (problems that interfere with producing a quality product on schedule and within cost constraints). Without focus on the right problems, quality circles often degenerate into meetings that accomplish little other than offering a forum for employees to complain about existing conditions. Without the benefit of skills available within the company but outside the immediate work area, quality circles often lack the orchestrated expertise required to solve most problems.

Litton Guidance and Control Systems in Woodland

Hills, California, was one of the first companies to recognize this shortfall in the quality circle approach. This organization is engaged in the development and production of ring laser gyroscopes, as well as other sophisticated electronics and electro-optical equipment. The company initially adopted the quality circle approach but found it lacking for many of the reasons described above. The company's president, Larry Frame, refined the quality circle concept into what Litton referred to as "focus teams." The Litton concept, under a variety of different names, quietly overtook the quality circle approach because it focuses the right people on the right problems. In many respects, the focus team approach is quite similar to quality circles, with two important distinctions:

- Management provides inputs to the focus teams to identify which problems it considers significant (instead of following a quality circle approach, in which shop floor personnel select problems to attack), and

- Individuals from other disciplines and organizations within the company, with the requisite skills to solve selected problems, are enlisted as focus team members (instead of only using personnel from the work center).

One might question if management participation in selecting the problems to be solved demotivates focus team members. The risk of doing so certainly is present, but Litton and other companies have not found this to be the case. Most team members realize the limitations associated with their own backgrounds, and soon come to welcome the expertise and advantages of a multidisciplinary problem-solving approach.

MIL-Q-9858A, MIL-STD-1520C, ISO 9000, and D1-9000 (Revision A)

Is the Litton concept brand new? Not really. Defense systems contractors will recognize that the focus team approach described here comes right out of MIL-Q-9858A (which is a government management standard on quality program requirements) and MIL-STD-1520C. ISO 9000 registered companies will similarly recognize the Management Review function, as will Boeing

suppliers working to D1-9000 (Revision A).

MIL-Q-9858A (*Quality Program Requirements*) requires systems contractors selling to the government to recognize adverse quality trends and to implement effective corrective actions.

MIL-STD-1520C (*Corrective Action and Disposition System for Nonconforming Material*) addresses the disposition and corrective action requirements associated with nonconforming hardware. This military standard requires a Corrective Action Board, which is an executive committee responsible for identifying and assuring that key quality issues are addressed. A defense contractor's Corrective Action Board is responsible for defining and assigning the quality improvement project teams to work these issues, reviewing progress, and providing management guidance.

ISO 9000 requires a similar function in its Management Review. ISO 9000 specifies that a supplier's management shall review quality records at appropriate intervals to assure that the quality system is effective. In practice in ISO 9000 organizations, in areas where the supplier's quality system is found to be less than adequate, teams are assigned to resolve quality issues.

Boeing's D1-9000 (Revision A) *Advanced Quality System*, which is literally written around ISO 9000, similarly incorporates requirements for the Management Review.

Back to Focus Teams

How does the focus team approach work? Figure 8-2 shows a concept that is working well in many organizations. We recommend creating a continuous improvement executive steering committee that includes the senior quality assurance, engineering, manufacturing, and procurement managers, and the company president. This leadership group can meet on a monthly or weekly basis to review quality measurement data, select key quality issues to be resolved, and provide a designated focal point for leading the effort and reporting on the focus team's progress.

Most companies base the selection of problems to be worked on two criteria: problems that

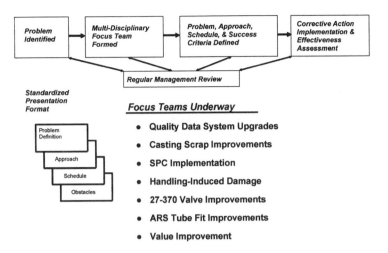

Figure 8-2. The Focus Team Approach. Many companies are abandoning quality circles and turning to this approach. Principal focus team advantages include focusing on the right continuous improvement challenges, and orchestrating the necessary skills to meet these challenges. An organization dedicated to continuous improvement will frequently have several focus teams underway simultaneously, all of which include work center members.

on the nature of the problem.

As was the case with quality circles, the focus team members must be provided with the tools necessary to solve the problems they are assigned. In many cases, these include training in problem solving and providing support from different organizations within the company on an as-required basis.

We recommend monthly focus team reviews for noncritical problems, and weekly reviews for more pressing problems. Once the right tools are provided, we believe the best thing that management can do to help the focus team is to leave it alone, except for reviewing progress on a regular basis.

immediately and severely impact the production of quality products on schedule are candidates for focus team assignment, and problems shown by the quality measurement system to be the most costly or most often occurring are also candidates for focus team assignment.

Team leaders should be selected based on their ability to understand the issue, their leadership and management skills (although the focus team leader need not be in management), and their objectivity. Objectivity is a critical ingredient, as vast amounts of time and money can be lost by investing in a focus team leader who refuses to keep an open mind or otherwise listen to the other members who make up the focus team.

The other members of the focus team can be decided by the focus team leader, or the company's executive quality committee, or perhaps both. We've observed team leaders recruited from the shop floor, quality assurance, engineering, manufacturing engineering, manufacturing supervision, and virtually all other internal organizations. In some cases, it might be appropriate to have the director of engineering serve as the focus team leader. In other cases, a lathe operator might be more appropriate. It all depends

Competently managed organizations dedicated to continuous improvement will have several focus teams underway at any given time (as implied by Figure 8-2). We recommend that focus team members adopt a simple four-chart approach for reporting progress. This standardized reporting format assures the inclusion of essential information, and allows for an organization's executives to review the activities of several focus teams in a one hour meeting. The four recommended charts are described below and shown in Figures 8-3, 8-4, 8-5, and 8-6. The charts shown in these illustrations were prepared for a focus team that attacked a material handling damage problem at a medium-sized aerospace supplier.

The first chart (see Figure 8-3) includes a brief statement of the focus team problem, the team members, a progress assessment, and quantitative success criteria. This first chart is critical. It forces a definition of the problem (which is the critically important first step in the four-step problem-solving process described in Chapter 5). The success criteria are also extremely important. Forcing this definition of success at the beginning of the focus team effort provides a quantified objective for the team members. It forms the basis for the focus team members to know if they have finished their job.

Problem Statement

Identify areas of high material-handling-induced damage, and reduce or eliminate such damage.

(Focus Team Members: Richards, McCloskey, Routt, Verett, Bonomo, Jakes)

Our Effort Is :

Succeeding

Holding On

Failing

Success Criteria:

Quantification of material-handling-induced damage by mid-December (by item and cost), and reduction of these values to less than 20% (of current value) by mid-February.

Figure 8-3. Recommended Focus Team Presentation: First Chart. This chart (prepared for a material handling damage focus team) identifies the problem and presents the focus team members, a progress assessment, and quantified success criteria.

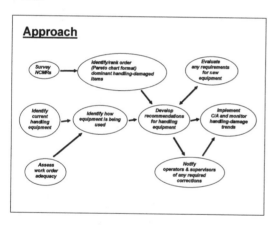

Approach

Figure 8-4. Recommended Focus Team Presentation: Second Chart. This chart shows how the continuous improvement challenge will be met.

Figure 8-4 shows the second focus team reporting chart, which presents a flow chart that shows the plan for solving the problem. The plan is based on the four-step problem-solving approach described in Chapter 5. Showing the plan in a flow chart allows both the team members and management to quickly determine if the approach is logical (i.e., if it makes sense and will lead to orderly progress in meeting the continuous improvement objective).

The third chart (see Figure 8-5) takes the plan

developed in the previous chart and shows it in a Gantt format (H.L. Gantt was an industrial engineer whose planning techniques, including the Gantt chart, were widely adopted during World War I and thereafter). Gantt charts are useful for rapid schedule evaluation. The Gantt chart lists all tasks, who is supposed to perform them, the start and completion dates, and whether the start and completion dates have been met. A Gantt chart also includes a "time now" line that readily shows if the project is on, ahead of, or behind schedule.

Schedule

Task	Name	Nov	Dec	Jan	Feb
NCMR Survey	Verett	■—■			
Pareto Chart	Richards	■—■			
ID Current Equip	Routt	■—□			
Assess Equip Use	Routt		□—□		
Assess Work Orders	Jakes		□—□		
Recommendations	Routt		□—□		
Notify Ops/Suprvrs	Bonomo		□—□		
ID Req'd New Equip	McCloskey			□—□	
Implement/Monitor	Bonomo			□—□	

Time Now

Figure 8-5. Recommended Focus Team Presentation: Third Chart. This chart shows the schedule in a Gantt format. Boxes at the start and completion of each task are filled in as the tasks are initiated and completed. The "time now" line shows any projects that are behind schedule. Note that the equipment usage task has not yet started, and is therefore behind schedule.

The fourth and final chart (see Figure 8-6) shows obstacles to progress, workaround plans for items behind schedule, and areas in which the focus team needs help. This is also a critical chart. Including the focus team's workaround plans is particularly critical, as this forces focus teams to zero in on anything that might impede meeting the quantified success criteria defined in the first chart. It helps the focus team members to recognize the issues management is likely to raise, and provides an opportunity to ask for help. By forcing recognition of these issues prior to focus team presentations, the focus team approach helps to develop the team members.

One of the key advantages to the four-chart presentation format described above is that it conveys the essential elements of information to management very quickly (this often takes less than ten minutes per presentation). The focus team approach helps to assure success in meeting the focus team's objectives, develops focus team members, and allows management to quickly review several focus team continuous improvement projects. Incidentally, the focus team working the material handling damage issue described in the preceding four charts greatly exceeded their pre-defined success criteria. Material handling damage was reduced to less than 5 percent of the mid-December values, and this reduction was attained earlier than the team's mid-February goal!

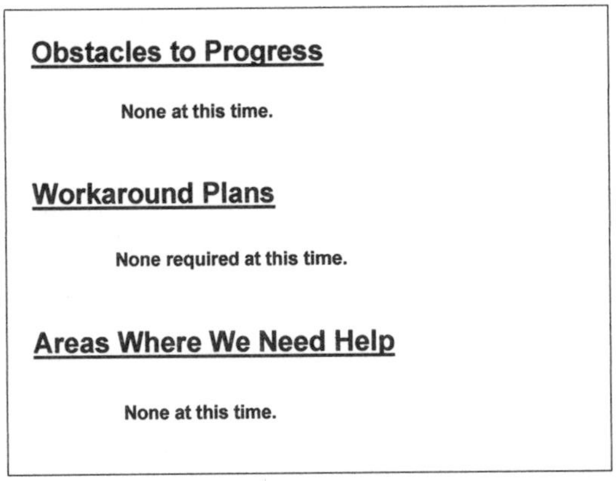

Obstacles to Progress

None at this time.

Workaround Plans

None required at this time.

Areas Where We Need Help

None at this time.

Figure 8-6. Recommended Focus Team Presentation: Fourth Chart. This chart shows obstacles to progress, workaround plans, and areas where help is required. An astute reader will note that the focus team members should have addressed the equipment usage task schedule slip identified in the previous chart. The equipment usage task is falling behind schedule, and a workaround plan to get back on schedule should have been included in this chart.

Summary

This chapter developed the concept of synergy being obtained by pooling employees' efforts to solve problems and meet continuous improvement objectives through the use of teams. Two teaming concepts were developed in this chapter: quality circles and focus teams.

Quality circles are groups made up of work center members chartered with implementing continuous improvement. There are certain disadvantages associated with quality circles, most notably an incomplete set of technical skills (not all skills necessary to solve a particular problem may be present in the quality circle team members), and lack of focus.

Focus teams are multidisciplinary teams chartered to focus on specific problems or continuous improvement objectives. Virtually every company in America is made up of groups of specialists; appropriately orchestrating these talents through a team structure has proven to be effective. Many organizations are finding the focus team approach to be more successful than quality circles in resolving problems and meeting continuous improvement objectives.

The above team-based problem-solving concepts, and focus teams in particular, are a natural fallout of the requirements emanating from MIL-Q-9858A, MIL-STD-1520C, ISO 9000, and Boeing's D1-9000 (Revision A) quality assurance standards.

References

Litton Industries TQM/EI Conference Presentation Materials, Simi Valley, California, August 29-30, 1990.

Total Quality Management, Part II: TQM Team Training, TRW Systems Integration Group Training Materials, January 1990.

"Speaking of Quality," *On Q, The Official Newsletter of ASQC*, February, 1993.

MIL-STD-1520C, "Corrective Action and Disposition System for Nonconforming Material," 27 June 1986.

MILQ-9858A, "Quality Program Requirements," 16 December 1963.

D1-9000 (Revision A), "Advanced Quality Program," Boeing.

ISO 9001, "Quality systems – Model for quality assurance in design/development, production, installation, and servicing," International Organization for Standardization, 15 March 1987.

Chapter 9

Statistics For Nonstatisticians

Thinking statistically...

Tom Lanson was confused. He examined the latest of seven generator Mk III trailers returned to Mobile Power Industries in the last week. The trailer's frame had fractured, as had the frames on the preceding six trailers returned to Mobile Power. The company had only delivered 400 trailers so far, and they were expecting to manufacture thousands. Having six returned with cracked frames was not a good sign.

Tom Lanson was Mobile Power Industries' project engineer for the company's latest mobile generator, the Mk III model. Mobile Power Industries manufactured towed generator systems, which consisted of small trailers with gasoline-powered generators, along with the equipment required to connect them to the tow vehicles (cabling, hitches, etc.). Mobile Power specialized in designing and manufacturing trailers to carry generators and gasoline engines purchased from other suppliers. The company integrated the trailers with the generators to provide a complete, ready-to-go generator system.

As the Mk III project engineer, Lanson had been responsible for designing and testing the trailer, which made the recent failures even more confusing to him. Lanson knew from the weight of the equipment the trailer had to carry (along with Mobile Power's previous experience in designing trailers) that the trailer frame would never see a load greater than 2,980 pounds. Lanson had designed the trailer frame to withstand 3,000 pounds to provide additional margin. He even had one of the frames tested to failure, and it indeed had failed at 3,000 pounds. Why were these frames failing?

Lanson initially wondered if the frames had been

abused, but there was no evidence of that. Each of the six cracked frames was from a different customer, so it was not likely that a single customer had abused the trailers. None of the six customers admitted to carrying additional equipment on the trailers, or to driving at high speeds on bumpy roads. Lanson even had one of the cracked frames examined by a metallurgist at a commercial failure analysis laboratory, suspecting there might be metallurgical defects in the frame material. The metallurgist told Lanson the frames rails contained no metallurgical defects. The frame rails had simply been overloaded and they failed, the metallurgist reported. The metallurgist suggested the design was inadequate for the weight of the equipment the trailer had to carry. Lanson had difficulty accepting that explanation. If the frames were overloaded, why weren't they all failing?

Deterministic Versus Statistical Thinking

Tom Lanson is a victim of his own deterministic thinking. What is deterministic thinking? Stated simply, it is forming absolute opinions (making a determination) based on a single observation. Lanson tested one trailer frame by loading it until it failed, and after observing that it failed at 3,000 pounds, he concluded that the frame would never fail at any load below 3,000 pounds. Based on his previous design and testing experience, he also knew that the loads experienced by the trailer would never exceed 2,980 pounds. Since 2,980 pounds is below 3,000 pounds, Lanson concluded that the frame should never fail and that his design was adequate. To many of us, making the above assumption seems logical. With a 20 pound difference between the load the trailer sees and the load it is capable of carrying, why should it fail?

Let's put this question in perspective. Consider the test Lanson performed, in which the trailer frame failed at a load of 3,000 pounds. Based on that test, Lanson felt comfortable that the trailer could withstand loads of up to 2,980 pounds (the maximum he predicted the trailer would ever experience). Lanson believed that a margin of 20 pounds (3,000 minus 2,980) was adequate.

Is a 20 pound margin adequate? Should Lanson have designed in a larger margin between the maximum load the trailer would ever see and its load bearing capability? Should he have designed the trailer to withstand 3,100 pounds, or 3,500 pounds, or perhaps even 4,000 pounds?

Perhaps Lanson could have headed in the opposite direction, and designed the trailer to withstand 2,990 pounds. This would have still provided a 10 pound margin between the heaviest load the trailer would experience and its capability. (With 20-20 hindsight and knowing that the trailers are failing, one is tempted to immediately say no, but let's assume for the moment that we are still in the design stage, and we don't know that the trailers are failing). Does a 10 pound margin seem reasonable? If it does, then let's make the question even more ridiculous. Why not have Lanson design the trailer to withstand 2,980.5 pounds? After all, the trailer will still be stronger (by a half pound) than the largest load it will experience. Or will it be?

To most of us, allowing only a half pound difference in the example above seems unreasonable. Ask why, and your reasoning will probably fall into one of two categories:

- What if the load on the trailer goes a little above 2,980.5 pounds?

- What if the trailer frame strength is a little lower on some trailers, and it can't withstand 2,980.5 pounds?

The above answers are intuitive. It just seems likely to most of us that nothing will be exactly repeatable. To put the problem in perspective, recall the experiment that Lanson performed when he loaded the trailer frame to failure and it failed at 3,000 pounds. As described earlier, Lanson then concluded that all of the trailer frames could

withstand 3,000 pounds. Ask yourself another question: If Lanson conducted the same experiment again, would the next frame fail at exactly 3,000 pounds, or would it fail at some value a little higher or a little lower than 3,000 pounds?

Does it seem intuitively unlikely that the next trailer frame would fail at exactly 3,000 pounds? Or does it seems more likely that the trailer frame would fail at a value near 3,000 pounds (either higher or lower than 3,000 pounds, but not exactly 3,000 pounds)? The answer to both questions is yes. Given that some spread will exist around the point at which the trailer frame is expected to fail (let's call this spread in the data its variability), is there a way to describe the variability such that we have a feel for its magnitude?

The Nature of Statistics

Statistics is a branch of mathematics used for describing or predicting observations. Webster provides a more exact definition:

Statistics. Facts or data of a numerical kind, assembled, classified, and tabulated so as to present significant information about a subject.

One of the problems with the above (and with classical approaches to statistics) is that they are intimidating. For now, let's simply accept that statistics are quantitative descriptions of how groups of things behave. To put this definition of quantitative descriptions in more understandable terms, suppose we have a room full of, say, 30 people. Without knowing anything else about this room full of people, we can make certain observations. We can examine the room to see if there are any midgets or giants. Assuming that not to be the case, we might develop a few preliminary, quantitative descriptions about the group:

- The weight of the people in the room probably varies between 130 and 230 pounds, and

- The height of the people in the room probably varies between about 4 feet, 6 inches and 6 feet, 6 inches.

The descriptions can be further refined by observing the weight and height of the people described above more closely. Perhaps most of the people seem to be about 5 feet, 9 inches in height (some are shorter, and some are taller, but the average height might be about 5 feet, 9 inches). Perhaps we further observe that most of the people appear to be of medium build, and therefore, we would estimate that the average weight might be about 165 pounds. Many of the people in the room might weigh more, and many might weight less, but 165 pounds seems to be a good representative weight.

Without going any further, we already have a set of statistics (or quantitative observations) to describe this group of people. The average height appears to be about 5 feet, 9 inches, and the spread around that average appears to vary from 4 feet, 6 inches to 6 feet, 6 inches. We similarly described the weight of the group. The average weight appears to be about 165 pounds, with a spread ranging from 130 pounds to 230 pounds.

Take a moment to think about the above quantitative descriptions. Doesn't it seem intuitive that such a description makes sense? At this point, however, our descriptions of the above group are strictly estimates based on visual observations. Suppose we wanted to more accurately define the weights and heights of the 30 people described above. To do that, let's actually weigh the people in the room and measure their heights. The data for these measurements are included in Figure 9-1, along with the average values of their weights and heights.

Most of us can readily accept the notion of an average value. In mathematical terms, the average is defined as the sum of all of the values divided by the number of values. Based on the actual data shown in Figure 9-1, we can calculate the actual average weight (as opposed to simply guessing at it, as done earlier) by summing all of the weights and then dividing this sum by the total number of people weighed. If all the weights are added together the result will be 5,026 pounds, and if this number is divided by 30 (the total number of people weighed), the answer is exactly 167.53 pounds. This number, 167.53, is the average weight of all the people in the room.

The same calculation can be performed for the average height. The sum of all heights is 1,997 inches, and when this is divided by 30 (the total

Person	Weight (pounds)	Height (inches)
1	142	62
2	228	77
3	176	70
4	193	70
5	166	68
6	152	64
7	169	68
8	193	72
9	158	66
10	178	71
11	156	62
12	148	60
13	131	56
14	169	69
15	225	75
16	142	60
17	171	70
18	138	52
19	145	61
20	172	70
21	133	57
22	180	72
23	165	69
24	169	70
25	196	73
26	134	58
27	163	67
28	158	66
29	210	73
30	166	69
Average:	167.53	66.57

Figure 9-1. Weights and Heights for 30 People. This table shows the heights and weights of 30 people, as well as the average weight and height for the 30.

number of people), the average height is found to be 66.57 inches, or just slightly more than 5 feet, 6 inches. Performing the above calculations yields exact values for the average weight and height for the group of 30 people described above. These are statistics that can be used to characterize the group.

Statistic	Value
Average Weight	167.53 pounds
Maximum Weight	228 pounds
Minimum Weight	131 pounds
Average Height	66.57 inches
Maximum Height	77 inches
Minimum Height	52 inches

Figure 9-2. Descriptive Statistics for Weight and Height. This table shows the spread in the heights and weights of 30 people.

What about the other numbers estimated earlier, the numbers that describe the spread of this data? Figure 9-2 shows that the upper and lower values for the group's weight are 228 pounds and 131 pounds, and the upper and lower values for the group's height are 6 feet, 5 inches (77 inches) and 4 feet, 4 inches (52 inches). These statistics further describe the group.

After making the above observations and calculations, we have a set of statistics that describe the group. Such statistics are called descriptive statistics. As we'll see later, once one has assembled a set of descriptive statistics, the data can be used to make predictions about the group (these statistics are called predictive statistics).

Suppose that in addition to simply knowing the descriptive statistics of average, minimum, and maximum values, one wants to know how close to the average most of the numbers lie. Stated differently, suppose we want to know how many weights are clustered near the average, and how many are near the extreme values. How can we get a feel for the relative distribution of these values?

Walter Shewhart's Head Sizes

During World War I, the United States faced many new challenges. One such challenge involved making large production runs of military equipment meeting stringent quality requirements. American military leaders recognized that traditional methods of manufacturing and controlling the quality of manufactured goods would be inadequate in light of the quantity of items that were needed by our military personnel, and they turned to Bell Laboratories for assistance in developing statistical techniques for quality control.

This task fell to Dr. Walter Shewhart, a mathematician at Bell Laboratories specializing in statistics. Dr. Shewhart applied his statistical knowledge to his first task, which was the design of a radio headset that would fit most troops. Dr. Shewhart recognized that in order to do this he needed to have a feel for the variability of the troops' head sizes, and he arranged to have the heads of 10,000 soldiers measured (which is not too different from what we did above when we recorded the weights and heights of 30 people). Dr. Shewhart and the Army measured 10,000 soldiers' head sizes and began to analyze the data. He found a peculiar characteristic, and that is that if the head sizes were plotted in a frequency distribution, the curve resembled a bell.

What is a frequency distribution? It's simply a plot of the number of times a point falls within a particular area. For the head-size frequency distribution, the data were arranged into equally spaced head-size groups between the highest and lowest head sizes. The number of times a head size fell into each group was counted, and then these numbers of head sizes in each group were plotted. The data might have appeared as shown in Figure 9-3.

Dr. Shewhart was intrigued by the bell shape of the curve, as it resembled similar curves he had encountered during other statistical studies. Shewhart subsequently found the bell-shaped frequency distribution described many other naturally occurring measurable events. This occurred so often that the bell curve frequency distribution was viewed as the normally expected distribution, and the characteristic bell curve became known as a normal curve. There are instances in which data do not adhere to the normal distribution, but these represent special cases, and we won't cover them here.

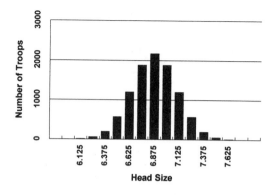

Figure 9-3. A Frequency Distribution of Head Sizes. Dr. Shewhart arranged 10,000 head sizes in a frequency distribution, which shows how many head sizes he found of each size between the smallest head size and the largest head size.

There are a couple of additional points we need to make about the normal curve. One normally begins construction of a frequency distribution by plotting values in their respective ranges, as shown in Figure 9-3. Once this is done, instead of representing the number of observations for any value by bars, we can instead plot a smooth curve along the tops of the bars. We can next say that the area under the curve is equal to one. Figure 9-3 has been converted in this manner, and it is shown in Figure 9-4. We'll see later in this chapter that this has important implications for how the normal curve can be applied for other purposes.

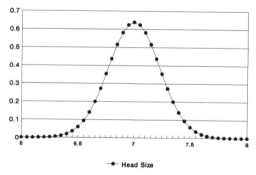

Figure 9-4. The Normal Curve for Head Sizes. The data shown in Figure 9-3 have been normalized so that the area under the curve is equal to one, and a smooth line has replaced the tops of the bars representing different head sizes. Presenting the normal curve in this manner will allow important applications in predictive statistics.

Will the weight and height data taken above (for our 30 people) follow a normal distribution? Let's

prepare a frequency distribution for weight first. To do this requires a graph with weight along the *x*-axis, and number of people in each weight category plotted along the *y*-axis. Figure 9-2 showed earlier that the weights of the 30 people range from 131 to 228 pounds, so the frequency might be prepared with weight ranges of, say, 10-pound increments from 130 to 230 pounds. This means that the frequency distribution will have weight ranges of 130 to anything less than 140 pounds, 140 pounds to anything less than 150 pounds, 150 pounds to anything less than 160 pounds, and so on, until we reach the last range of 220 pounds to anything less than 230 pounds. (Note that when we specify the weights from one value to just under the lower weight of the next range, we do so to define in which group any weight that is exactly equal to the boundary value should lie.) An examination of the weights of the 30 people shown in Figure 9-1 will show the number of people in each 10 pound weight category. These data are summarized in a tabular manner in Figure 9-5.

Weight Range	People In Each Range
130 to just under 140 pounds	2
140 to just under 150 pounds	4
150 to just under 160 pounds	4
160 to just under 170 pounds	7
170 to just under 180 pounds	4
180 to just under 190 pounds	1
190 to just under 200 pounds	2
200 to just under 210 pounds	0
210 to just under 220 pounds	1
220 to just under 230 pounds	2

Figure 9-5. Frequency Data for Weight. This table shows the frequency of weights in each ten pound range for the 30 people described in this chapter.

The last step in preparing a frequency distribution is to show the data on an *x-y* plot. The data shown above are shown in the Figure 9-6 histogram.

We can also prepare a frequency distribution for height. If we take height in 4-inch increments from 50 to 78 inches, we have comparable data shown tabularly in Figure 9-7 and graphically in Figure 9-8.

Number of People in Each Range

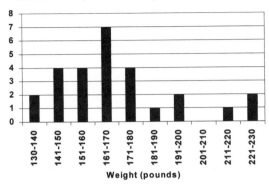

Figure 9-6. Frequency Distribution For Weight. This frequency distribution graphically portrays the data of Figure 9-5.

Height Range	People In Each Range
50 to just under 54 inches	1
54 to just under 58 inches	2
58 to just under 62 inches	4
62 to just under 66 inches	3
66 to just under 70 inches	8
70 to just under 74 inches	10
74 to just under 78 inches	2

Figure 9-7. Frequency Data For Height. This table shows the frequency of heights in each four inch range for the 30 people described in this chapter.

Number of People in Each Range

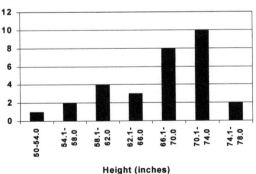

Figure 9-8. Frequency Distribution for Height. This frequency distribution graphically portrays the data of Figure 9-7.

As can be seen in Figures 9-6 and 9-8, the data approximate a normal curve, but only very loosely. They do not closely match the smooth shape of Shewhart's head size bell curve, as shown in Figure 9-3. Why is that?

The answer is that there will always be inherent variability in the data, and when plotting a small number of points, we can expect considerable departures from the bell curve. The more points we plot in the frequency distribution, though, the more we can expect the curve to follow the classic bell shape. A quick way to convince yourself of this is to plot only two points on the frequency distribution, and note that the result looks nothing like a normal distribution. Plot two more, and the result is the same. Plot 10 data points, and we may see a pattern begin to emerge. Plot all 30 data points, and the data start to approximate the bell pattern. If we were to plot 300 data points on each of our weight and height curves, we might see the curve more closely adhere to the classic bell pattern. When Shewhart plotted 10,000 points, it's likely he saw a near-perfect bell shape.

The Mathematics of the Normal Curve

The normal curve can be described mathematically with the following equation:

$$y = \frac{e^{-(1/2)(x-u)^2}}{\sigma\sqrt{2\pi}}$$

where:

y = the number of times a value occurs in the distribution

e = 2.7182, which is a constant that is the base of the natural logarithm

x = the value whose frequency we are interested in finding

υ = the average value

σ = the standard deviation

π = 3.178, which is another constant

Figure 9-9 shows the concept. At first glance, the above equation looks intimidating, but it really isn't. For our purposes, we can simply say that it represents a way of determining the shape of the frequency distribution based on two factors that vary and several other things that do not (the things that do not are constants). Note that in the above equation, all of the terms except two are constants. The only two factors in the above equation that are not constants are the average *x* value (which is represented by the symbol υ) and the standard deviation (which is represented by the Greek letter σ).

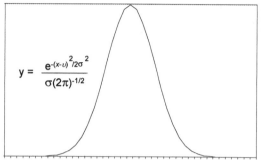

$$y = \frac{e^{-(x-\upsilon)^2/2\sigma^2}}{\sigma(2\pi)^{-1/2}}$$

Figure 9-9. The Normal Distribution. Once the standard deviation and average (or mean) values are known, one has fully described the distribution.

What is the symbol υ in the above equation? The symbol υ simply represents the *x*-coordinate upon which the bell curve is centered. In other words, the top of the bell curve is directly over υ, and because of that, we know that υ is actually the average value of the parameter being plotted. That's because the center (or top) of the bell curve typically represents the value about which all other values are centered.

The other factor in the above equation that is not a constant is the standard deviation, which is represented by the Greek letter σ. The standard deviation is significant because it represents the bell curve's spread, or stated differently, the spread in the data about the average.

Figure 9-10 shows two bell curves with different standard deviations. A frequency distribution with a small standard deviation represents a sharply defined bell curve, while a frequency distribution with a large standard deviation represents a more flattened bell curve. What that means to us is if the standard deviation is a relatively small number, the variability in the data about the average is small (the data are more nearly clustered near the average). On the other hand, if the standard deviation is a relatively large number, there is more variability in the data, and the data will be spread out more (even though it will still be centered about the mean).

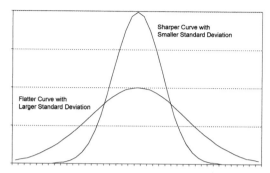

Figure 9-10. Bell Curves with Different Standard Deviations. The sharper curve has a smaller standard deviation, and because of this, there is less spread in the data about the average value. The flatter curve has a larger standard deviation, and its spread is larger. This means that the data are less tightly grouped around the average.

How does one determine the average and the standard deviation? These calculations are fairly straightforward. As we defined earlier, the average is simply the sum of all values divided by the number of values (we performed these calculations when considering our population of 30 peoples' heights and weights). Calculating the standard deviation is only slightly more complex. Here's the formula for calculating the standard deviation:

$$\sigma = \sqrt{\frac{\Sigma x^2 - n(\upsilon)^2}{n-1}}$$

where:

σ = the standard deviation
x = the value of each sample
Σx^2 = the sum of the squared values for all of the samples
n = the total number of values
υ = the average value

What does the above tell us? If we have a set of data (such as heights, or weights, or head sizes, or trailer frame breaking strengths), it tells us a great deal. It tells us we can compute the average value, and the standard deviation, and thereby derive a feel for the average value and the spread of the data about this average value. This is significant, because it tells us that if we know the average and the standard deviation of a group of data points, we can define the normal curve for that distribution. Why is that important? One reason is revealed in another name for the normal curve, and that name is the probability distribution function.

Implications of the Normal Curve

Let's reconsider the nature of the normal curve and what it represents. Earlier, we developed the concept of a frequency distribution, and we showed that frequency distributions represent how many times a value will lie within a specified range. Once the frequency distribution is known, therefore, it can be used as a tool for predicting how often a value will be within a certain range (rather than simply describing how often previous observations were within the range).

The above observation is significant from several perspectives. First, it forms the basis for a line between everything discussed thus far and most of what we are about to do. Up to now, our statistics discussions have focused on using statistics to *describe* observations made in the past (either during a test, or by collecting inspection data, or perhaps by examining other quantitative data). Our discussion will now turn to techniques to use statistical data to <u>predict</u> future outcomes.

Returning to Mobile Power Industries

Let's return to Mobile Power Industries' trailer dilemma, and examine how Tom Lanson might have approached his problem differently. Recall that Lanson had performed testing during the trailer design effort in which he tested one trailer to failure and observed that it failed at 3,000 pounds. Lanson also knew (or so he thought) that the trailer would never be loaded with more than 2,980 pounds. In light of the discussion in the preceding pages, we can now recognize that Tom Lanson did not have a feel for the distribution of trailer frame strengths. All he knew is that one trailer failed at 3,000 pounds. He didn't know if the next one would have failed at 2,900 pounds or at 3,500 pounds. Lanson did not know the mean or the standard deviation for the trailer frame's strength distribution. Lanson used a single data point (in this case, one measurement of trailer strength) to make a determination about the strength of all the trailer frames. He was a victim of his own deterministic thinking.

What should Lanson have done? Suppose Lanson tested several frames to failure, and based on the results of this test he calculated the average and the standard deviation of the trailer frame distribution strength using the formulas provided earlier in this chapter. If Lanson compared these numbers to the maximum load the trailer frame would have to withstand, he would then have a feel for how often the trailer frame would fail. We'll see how to make this determination in a few more paragraphs.

The important thing to recognize at this point is that the trailer frames are not all going to fail at exactly 3,000 pounds. Some will be higher, and some will be lower. If Lanson recognized this earlier, he might have recognized that based on the average trailer frame strength and its standard deviation a certain percentage of the trailer frames would have strengths less than 2,980 pounds. That would explain why so many are failing, even though one trailer frame (the one Lanson tested when designing the trailer frame) failed at 3,000 pounds.

Let's assume that Tom Lanson tested a number of trailer frames to develop the frequency distribution described above. If Lanson knows the mean and standard deviation of this frequency distribution, he will then know the average strength of the trailers, and the spread of strengths around this average. Let's say that the mean value found in Lanson's testing showed that the trailer's average strength was 3,000 pounds, and the standard deviation was 13 pounds. Figure 9-11 shows this normal curve. Knowing the mean and standard deviation for trailer strength, let's ask the following question: What is the probability that a trailer will fail at or below a particular value (in this case, 2,980 pounds)?

In order to answer this question, we need to consider another very important characteristic of the normal curve. As mentioned earlier, the normal curve represents a probability density function. What that means to us is the normal curve represents a population, and from it, we can determine the likelihood that a particular sample will lie above or below a value on the *x*-axis. This determination is based on how many standard deviations a value is from the mean of the normal curve.

Here's how this concept works. A characteristic of the normal curve is that about 68 percent of the area beneath it lies within one standard deviation on either side of the mean (this concept is shown in Figure 9-12). This tells us that the probability of being within one standard deviation from the mean is about 68 percent. It also tells us that the probability of a value being more than one standard deviation away from the mean (on either side of the mean) is only 32 percent (or 100 percent minus 68 percent, since we are working with 100 percent of the area beneath the curve). If we want to know the probability of a value being more than one standard deviation below the mean, we simply look at half of the distribution, and we see the probability is half of 32 percent, or 16 percent. We can do the same thing for two standard deviations. The probability of being more than two standard deviations away from the mean (on either side of the mean) is only about 5 percent. The probability of being more than three standard deviations away from the mean (on either side of the mean) is only about 0.3 percent.

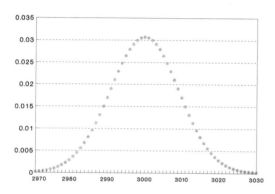

Figure 9-11. A Normal Curve for Frame Strength. The mean value is 3,000 pounds, and the standard deviation is 13 pounds.

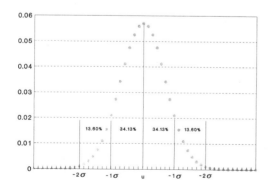

Figure 9-12. Using Normal Curves to Predict Probabilities. The standard deviation concept can be used to predict the likelihood of values being greater or less than selected points on the normal curve.

How can we use the above information to determine the likelihood of being above or below a certain value? If we can express the distance from the mean in terms of standard deviations, we can use the probability distribution characteristics of the normal curve (as explained above) to determine the probability of being above or below any value. If the value we select is exactly one or two or three standard deviations away from the mean, then determining the probability is as straightforward as explained above. What happens, though, when the value is not an integral number of standard deviations away from the mean? The easiest way to do this is to convert the distance from the mean to the number of standard deviations, and use statistics tables based on the normal distribution to make this determination. These tables are found in any statistics textbook, and are based on the principles described above.

Let's return to the normal curve in Figure 9-11 and ask the question: How likely is it that a trailer will fail at a value below 3,000 pounds? By examining the curve, we can see that this probability is equal to 50 percent (since the mean value is exactly 3,000 pounds, half of the trailers will fail at values higher than 3,000 pounds, and half will fail at values lower than 3,000 pounds).

So far, this is similar to the discussion above. This is almost too straightforward, in that selecting the mean value for trailer strength made things easy for us. What happens if we pick a value somewhere else on the trailer strength normal curve? Suppose we want to know what percentage of trailers will fail at values of 2,980 pounds or less? How do we make this determination?

The above question is where we really start to rely on mean values and standard deviations. If we know that the mean trailer strength is 3,000 pounds and the standard deviation is 13 pounds, we can determine how far away from the mean our selected value of 2,980 pounds is, and express this as a number of standard deviations. Here's how we do this:

- The difference between 3,000 pounds (the mean) and 2,980 pounds (the value of interest) is 20 pounds.

- The 20 pounds can be expressed as 20/13 standard deviations, or 1.54 standard deviations.

Knowing that the trailer strength of concern is 1.54 standard deviations away from the mean, we need merely refer to a table for the normal curve in any statistics text to determine what area under the curve is left when a point is more than 1.54 standard deviations away. The normal curve table tells us that this remaining area is 0.012, which is equal to the probability of the trailer strength being at or below 2,980 pounds. In effect, it tells us that if the trailers are all subjected to a load of 2,980 pounds, we can expect about 1.2 percent to fail.

What does the above mean to Tom Lanson and Mobile Power Industries? It means that he needs more margin between the strength of the trailer frame and the largest load it will experience. The trailer strength has a distribution, and Lanson's challenge is either to shift the mean of this distribution further away from the maximum load the trailer will experience, or tighten the standard deviation such that the probability of the maximum load exceeding the strength is acceptably low. We'll see how this is done in subsequent chapters.

Summary

In this chapter, we developed the concepts of descriptive and predictive statistics, the normal curve, and averages and standard deviations. Descriptive statistics are used to quantify our observations about groups of things. The normal curve represents a frequency distribution of data about an average value. The standard deviation represents the spread of the data about the average value. Knowing the average value and the standard deviation allows us to make predictions about future outcomes. These are basic statistical tools upon which many concepts in the following chapters will be based, including statistical process control, analysis of variance, and Taguchi testing.

Thinking statistically is a basic part of quality management philosophy in a high technology manufacturing environment. It helps prevent nonconformances by recognizing the variabilities in processes, and in product strength and load distributions. Much of the challenge in implementing continuous improvement is centered on characterizing these distributions to predict when processes are drifting out of control (and taking steps to bring them back into control prior to making nonconforming hardware), and in minimizing variability.

References

How to Use (and Misuse) Statistics, Gregory A. Kimble, Prentice-Hall, Incorporated, 1978.

Statistical Quality Control Handbook, AT&T, Delmar Printing Company, 1956.

Probability and Statistics for Engineers, Irwin Miller and John E. Freund, Arthur D. Little, Incorporated, 1965.

Quantitative Techniques for Management, Thomas A. Payne, Reston Publishing Company, Incorporated, 1982.

Chapter 10

Statistical Process Control

A secret weapon we gave to Japan...

Arleigh Henderson was perplexed. He stood in front of the F-16 fuel tank at Sargent-Fletcher's test stand looking at yet another tank that had failed the shutoff capacity test. Sargent-Fletcher, as the original developer and only producer of the F-16 externally mounted wing tanks, had built over 20,000 of the aluminum tanks since the early 1980s. Sporadic failures for shutoff capacity had plagued the company during the entire production run. Henderson thought that a company in business that long, and one building the same product, ought to be able to make tanks without recurring failures. As a new quality engineer assigned to the program, Henderson was now responsible for dispositioning all tank failures, and it bothered him that the failures had not previously been eliminated.

"Can you tell me exactly what happened?" Henderson asked Boyd Braddock, one of the test technicians.

"Sure, nothing new here," Braddock answered. "The tank shut off with 375.4 gallons of fuel in it. Maybe every month or so we get a run of tanks that come in either low or high. Sometimes they shut off too early, and sometimes they shut off too late."

"What's the tank supposed to do? I mean, can you tell me what this test is for?" Henderson asked Braddock.

"Well, we begin by mounting the tank on the test stand and performing a visual inspection," Braddock answered. "If there are no dings or scratches on the tank, and the labels are all properly prepared, we then move on to the performance portion of the acceptance test. We test for a lot of things, like weight, center of gravity, differential pressure,

drainage, and so forth. One of the tests, the one we occasionally fail and the one we're out here talking about now, is the shutoff capacity test. To do this test, we pump fuel into the tank with the test stand plumbing, starting with an empty tank. The plumbing and valves in the tank are supposed to let the tank fill up and then shut off. The shutoff feature is supposed to occur when the tank has between 370 and 374 gallons. Anything below 370 gallons, or above 374 gallons, and the tank fails."

"Does it really mean anything to the airplane, the F-16, if that happens?" Henderson asked.

"Probably not," Braddock answered, "but I'm really not the right guy to ask. My guess, though, is that if the tank really only had 369.5 gallons the airplane would never know the difference. Same goes for 375 gallons. That's not the point, here, though. The Air Force wants the tank to hold between 370 and 374 gallons when it's full, and that's what they're paying us to build for them."

"You're right," Henderson said. "How often do these failures occur?"

"We might see one or two a month," Braddock said. "Sometimes we'll see six or seven in a row. Sometimes they're too high, and sometimes they're too low. When we get a group that fails, though, you know, four or five in a row, they all seem to be either on the low end of the specification or on the high end. When they fail in a group they don't go low and high. They cluster on one side of the allowed tolerance band."

"I see," said Henderson. "What about before they fail? What do you see then?"

"Well, we don't keep records like that," Braddock answered. "We just record whether they pass or fail. We don't record the actual values, even though the flowmeter tells us the actual value, and it does so to a tenth of a gallon."

"Any reason why not?" Henderson asked.

"That's what the test data sheet asks for, you know, just pass or fail, no values or anything," Braddock said. "But since you asked, I can tell you that we generally know when one is going to fail. Not always, but a lot of times we'll see the tanks start to go up or down."

"How do you mean, go up or down?" Henderson asked.

"We'll see the tanks' shutoff values trend up or down," Braddock answered. "We'll see a tank at, say, 373.0, then one at 373.4, then one at 373.6, then one at 373.8, you know, with the shutoff capacities moving up. Then we'll get one that goes over the magical 374 gallon limit. The same kind of thing happens when they go below 370 gallons. It's no surprise to anyone who's worked out here for a while. You don't have to be a genius to see a failure coming when they start to trend up or down like that..."

The Dangers of Driving Blindfolded

"You don't have to be a genius to see a failure coming." Boyd Braddock's words were more insightful than perhaps either he or Arleigh Henderson realized. Before launching into a discussion of one of the cornerstones of a successful continuous improvement concept, statistical process control, let's briefly consider our earlier discussions of the prevention versus detection quality management philosophies. You should recall from our earlier discussion that one approach to quality management involves building product and then inspecting it to sort good from bad. This is the classical detection approach (i.e., relying on inspection). The prevention approach takes another tack, and seeks to prevent defects from occurring.

There are a multitude of tools available for preventing defects (as presented throughout this book). One of the most powerful, though, is statistical process control, and one of the principal advantages of statistical process control is that it places the responsibility for quality squarely in the hands of the operator (not a downstream inspector who looks at the product after it is built).

How does this process of placing responsibility for building good product in the hands of the operator work? Before continuing, let's consider an analogy. Most of us drive automobiles. Think about what happens when you are driving on the freeway. You know how fast to drive to keep up with the flow of traffic, and you know how to steer the automobile to keep it centered between the painted lines denoting your lane. For most of us who have been driving awhile, this occurs almost automatically. The process of driving requires that we know the boundaries in front of, behind, and to the left and right of our car. We don't need anyone else in the car to tell us when to slow down or speed up, or to stay closer to the left side or right side of our lane. In fact, most of us resent it when one of our passengers (or perhaps a fellow motorist in another car) attempts to provide such direction. You are driving the car. You know how to keep it in control, with "in control" meaning the limits defined by the flow of traffic and the lane's boundaries.

Now, let's imagine something very different when you drive the same automobile on the same freeway. Suppose this time you find yourself moving at freeway speeds in traffic, but we blindfold you and put a fellow passenger in the car to tell you when to speed up, slow down, veer left, or veer right. Could you drive a car this way?

Perhaps, but even the thought of attempting to operate a complex piece of machinery in the manner described above is, at best, frightening. Let's take it a step further, and forbid your fellow passenger (the one telling you to steer left or right and how fast to drive) from telling you anything until you have either crossed into another lane or crashed into another car. Driving in this manner effectively guarantees an accident.

Think about the above.

Now, think about companies that rely on inspection as a means of assuring quality, and how they operate. The people building product don't know if the hardware they are building is good or bad (they're blindfolded as far as this information is concerned, much the same way as you were when driving your car in the example above). After these blindfolded operators build product, they give it to an inspector to determine if it is good or bad (much the same as you relied on your fellow passenger for driving feedback, but with the added restriction that you could only get feedback after leaving your lane or crashing into another car). When considered in this light, is not relying on inspection an inefficient means of creating a quality product? Would it not be better to take the blindfolds off of the operators, and allow them to see where they are going in their desire to build a quality product? The answer to both questions, of course, is yes. Statistical process control is the tool that removes operators' blindfolds, and in so doing, moves organizations from a detection-oriented quality management philosophy to a prevention-oriented philosophy.

Industrial Revolution to the Information Age

Many of us, when faced with examples of poor quality, often hear comments such as "there's just no pride of workmanship anymore." The implication of these sorts of comments is that quality deficiencies are often due to workmanship, or operator errors. Modern quality management maintains that very few nonconformances are due to worker error, but rather, most are due to defective manufacturing processes, poor management systems, and poor product design.

We agree totally with this concept. Our experience in every company we've visited and our readings have convinced us that it is true. Most quality deficiencies are not due to workmanship errors. But let's turn back to the "no pride of workmanship" train of thought. Where did it originate?

Early in our nation's history, most items were individually made by craftsmen. A chair, for example, would be produced by a single individual, who built the item from raw lumber and turned it into a finished product. Clothes were produced the same way. So were guns, wagons, and virtually every manufactured product. Techniques of mass production had not yet arrived, so a single individual did in fact have responsibility for building an item from start to finish. Any mistakes in its fabrication

were exclusively his or hers. Individuals recognized this, frequently took great care in creating an item, and did indeed have "pride of workmanship."

Pride of workmanship notwithstanding, there were major problems with the above approach. One was that the process of creating an item took a great deal of time. Because of that, items tended to be more expensive than they had to be. Yet another problem was that when an item broke, especially if it happened to be something as complex as, say, a gun, getting it repaired was strictly a custom proposition. One could not simply buy a replacement part and drop it in. That's because of the "craftsmanship" involved in individually hand fitting each piece.

Mass Production Emerges

The industrial revolution, about which we'll say more in a minute, largely began in the mid to late 1800s. Even before that time, Eli Whitney (inventor of the cotton gin) recognized that in order to meet the needs of the emerging American mass market one needed an efficient manufacturing process that could produce identical items in large quantities at a reasonable cost. In the late 1700s, Whitney believed that if he could standardize component dimensions for the parts used in complex mechanisms, he could achieve the goal of parts interchangeability, and build items in mass production. Whitney reasoned that if he could produce large numbers of each of the components used in a complex mechanism (and make all of them identical), then workers could simply assemble the parts to create the complex mechanism.

Whitney sold his idea to the United States government when he accepted a contract to build 10,000 rifles for the Army using this concept of mass production and parts interchangeability. The idea was sound, but it was too far ahead of its time. Whitney began the contract by making enough parts for the first 700 rifles. With 700 complete sets of parts, Whitney found (much to everyone's dismay) that he could only assemble 14 rifles.

What went wrong? Using the manufacturing processes of the era, the various rifle parts could not be made such that they were identical to each other, and therefore, they were not completely interchangeable. The concept of component interchangeability was sound, but the engineering of component tolerances and the ability to hold the tolerances necessary to achieve interchangeability had not yet arrived. The parts had too much variability.

Scientific Management

Whitney's concept of mass production did not get off the ground until the later 1800s (after the U.S. Civil War), when the industrial revolution began in earnest. Frederick Taylor, regarded as the father of modern industrial management, put forth the notion of scientific management of work. Scientific management gained great favor in the United States and Europe. Taylor's work emphasized the division of labor, much the same as had Whitney's earlier ideas. Taylor recognized the economies of scale associated with dividing the tasks necessary to make a finished product and apportioning these tasks amongst the workers in an organized factory.

Unlike his predecessors, however, Taylor also recognized that this new concept of division of labor brought with it new burdens. One was the absence of the craftsman-like approach in which a single worker takes a product from raw material to finished item, and its inherent pride of workmanship.

Taylor saw that division of labor necessarily created dull and repetitive tasks for factory workers, and that methods of motivating workers to maintain high output would be necessary. This fact notwithstanding, Taylor seized upon time and quantity as the fundamental standards for measuring worker output in this new environment. Worker adequacy would be judged by ability to meet a specified production quota.

Although there are many advantages associated with scientific management of work (the philosophy is often referred to as Taylorism), there are also significant disadvantages. Taylorism introduced two concepts that work against quality.

The first detractor is the division of labor concept. In addition to dividing the tasks required to produce a finished item, Taylorism frequently resulted in separate inspection functions. For the first time in American industry, people responsible for performing a task were no longer responsible for the quality of the task they performed. Someone else worried about that (these people became known as inspectors, and the organizations to which they belonged became known as quality control).

The second Taylor-induced shortfall was judging peoples' output solely on rate. How well an item met requirements no longer mattered. How many were produced per hour did.

Taylorism rapidly gained an intrenchable foothold in this country and other industrializing nations. Even today much of our manufacturing management philosophies are dominated by concepts firmly rooted in Taylorism (if you doubt this, try to find a factory that doesn't rely on manufacturing standards or specified time periods to produce an item).

SPC Stirrings in America

The beginnings of change began to emerge in the 1920s. Recall our earlier discussion in the last chapter about Walter Shewhart's work in designing a headset for U.S. Army radio operators. Shewhart, at Bell Laboratories, continued his work with statistics and recognized that just as peoples' head sizes varied, so too did the dimensions of items manufactured using mass production techniques.

Shewhart recognized that the variability associated with manufactured component dimensions followed a normal distribution, and he reasoned that one could therefore track component dimensions to determine if they were starting to drift out of the normal range. Shewhart recognized that as long as the dimensions remained in the normal range, their variability was under control. If they started to drift out of the normal range, something special was occurring to take the process producing the parts out of control. If one could see this departure beginning to emerge by tracking dimensions, one could then introduce corrections very early, perhaps even early enough to prevent making any parts that were outside allowed tolerance limits.

Shewhart refined his approach and published a book (*Economic Control of Quality of Manufactured Product*) in 1931, but his concept of statistical process control did not initially gain favor or widespread acceptance. Statistics, if not treated properly, is a fairly intimidating subject even today.

In 1931, manufacturing management just wasn't ready for it.

Sampling Plans Gain Favor

Managers of the Shewhart era recognized, however, that some form of reason was needed to simplify and provide structure to the statistics of the inspection process. The challenge at that time was not perceived to be one of preventing defects, but was instead viewed as how to economize on the inspection function. Managers wanted a way to intelligently select samples from their lots of components and finished products to determine their acceptability, instead of having to inspect every single component or finished product.

Around the same time that Shewhart published his work on statistical process control, two other men at Bell Laboratories, H.F. Dodge and H.G. Romig, were developing standardized sampling approaches to aid in the inspection process. Dodge and Romig offered a sampling approach that had a solid mathematical framework, was fairly simple to comprehend, and offered a more economical way to perform the inspection function.

This Dodge-Romig approach was based on examining a sample pulled from a production lot instead of the entire lot. To manufacturing managers of the 1930s, it was exactly what they thought they needed, and the Dodge-Romig statistical sampling approach gained widespread acceptance. Unfortunately, it also put this country in a more entrenched detection mode of doing business. Inspection became the norm. Sorting bad product from good became a way of life in industrial America. Inspectors flourished, as did those who bought and sold scrap material.

There were and are significant shortfalls to the inspection sampling approach. One is that the entire lot is usually finished before the sample is pulled, and if it is bad, an entire lot may have to be scrapped if the lot cannot be salvaged. Even if the lot can be reworked or otherwise salvaged, the cost of quality goes up. Another disadvantage is a concept euphemistically called consumer risk, and that is that even though the sample pulled from the lot passes its acceptance test, the lot may have defective parts in it.

Statistical Process Control Helps the War Effort

World War II began for the United States in 1941. All kinds of complex items were being made in enormous quantities to support the war effort. This was particularly true in the ordnance industry, where bombs and bullets were being made by the millions and billions. A few insightful people recognized that simply sampling and performing inspections on munition lot samples would not work under these circumstances. The cost of rejected lots of ammunition would be enormous in financial terms, but the dollar loss would be trivial compared to the cost of not providing ammunition to military forces that desperately needed it (or worse yet, providing munitions that did not work).

Clearly, the United States needed a management approach that would reduce scrap levels and assure that only good product reached our troops. A preventive approach was needed. Walter Shewhart's work came to light again, and for the first time, the United States adopted statistical process controls on a large scale. This first major application involved the United States' wartime munitions manufacturing facilities.

Curiously, when World War II ended, so did America's interest in statistical process control. Munitions manufacturing facilities scaled back enormously or disappeared altogether. Those that remained returned to lot sampling methods of quality control. Statistical process control had not caught on elsewhere in American industry, and to a great extent, the concept died in this country.

Japan Accepts What America Rejects

At the close of World War II, Japan was a devastated nation. The Japanese industrial base was effectively destroyed, the government no longer functioned, and basic human needs could not be met. The United States installed General Douglas MacArthur as military governor of Japan immediately after the war, which was a position that gave him far-ranging powers and authority.

MacArthur recognized that one of his immediate priorities was to help the Japanese to begin rebuilding. MacArthur knew he was a military man, and he needed help to guide the Japanese in rebuilding their industrial capabilities. He enlisted the aid of a relatively obscure (at the time) American management consultant named W. Edwards Deming.

When Deming came to help the Japanese, he found a remarkable situation. The Japanese people had no remaining industrial culture. Their industrial base had been eliminated, and Deming found an audience willing to listen to anyone who might help. The Japanese were no longer burdened with an inspection philosophy of quality management. They had no management philosophy at all due to their post-war environment. Deming had a clean slate upon which to draw his management doctrines.

It's important to recognize that Japan (both before and after World War II) is a small island nation with no natural resources except the Japanese people's intelligence and industriousness. The Japanese then, as now, were a frugal people. Japan's lack of natural resources made the Japanese people especially receptive to concepts that maximized use of imported materials. A management philosophy that espoused minimal waste was particularly attractive to a country that had to import virtually all of its raw materials. Deming's philosophy offered just such an approach. Deming was a believer in statistical process control.

The Japanese took to Deming's teachings with fervor. The results were not immediate, but Deming instilled in the Japanese an underlying statistical approach that would offer strong worldwide quality and marketing advantages in coming decades.

Maintaining A Detection Mindset

Oblivious to the emerging quality of Japanese products, post-World War II American industry remained firmly entrenched in an inspect and detect quality management philosophy. Americans were caught in the post-war boom created by returning soldiers, sailors, and marines who were making up for a half decade of delayed purchases. Goods were selling, and selling well. American industry was widely regarded as the best in the world. Everyone in America knew that anything made in Japan was junk. There seemed to be no need to minimize scrap or undertake seemingly complex statistical process control approaches to manufacturing management.

Why change when life was so good?

All the while, Deming's teachings on statistically based quality management continued to take root in Japan. Our first inklings of Japan's looming emergence occurred when they began to make significant inroads into the American consumer electronics market in the mid-to-late 1960s. Another early indication occurred when Honda began to decimate the international motorcycle market with their "You meet the nicest people on a Honda" advertising campaign, backed up by inexpensive, high reliability, and high quality motorcycles.

In 1973 something else happened that changed America. Oil embargoes hit the United States, and suddenly, the gas-guzzling monsters American automobile manufacturers had offered for decades began to lose market share to smaller and more economical foreign automobiles. Cars like Hondas, Toyotas, and Datsuns impressed American consumers with their quality, reliability, and low cost. The oil embargoes ended, but Americans had experienced high quality, and they weren't about to give it up. American automobiles lost market share steadily through the 1970s, even after the long lines at gas stations had disappeared. The American consumer had been introduced to Japanese quality in a big way, and suddenly, the relatively shoddy workmanship that seemed to be inherent to American automobiles was simply no longer acceptable.

What did the Japanese have in automobiles, consumer electronics, cameras, and motorcycles that American industry did not? In a word, quality. American consumers recognized that Japanese products more nearly met their needs and expectations. The fit, finish, and inherent quality in Japanese products was irresistible for many of us. One of the key Japanese management tools in attaining these attributes was (and still is) statistical process control.

The Sleeping Giant Awakens Once Again

American industry began to recognize the extent of its problem as we entered the 1980s. Automobile manufacturers had lost major portions of their market share to the Japanese. Harley-Davidson, the only remaining American motorcycle manufacturer, had become a minor player in an industry it had previously dominated. Most of the American camera market belonged to the Japanese. Most American consumer electronics companies (companies that had dominated their markets) either lost significant market share or ceased to exist.

A curious phenomenon occurred when American industry began the struggle to turn the situation around. The Japanese had previously turned to America for industrial management guidance, but America now sent people to Japan to study why Japanese manufacturing methods worked so well. We discovered many concepts underlying Japanese quality (indeed, many of the chapters in this book discuss quality concepts that originated in Japan). We found that one of the dominant reasons for Japan's quality superiority was their use of statistical methods for controlling manufacturing processes. The Japanese success story was rooted in the teachings of an American, W. Edwards Deming, a man whose teachings never caught on in America. The result of this discovery? American industry began a wholesale movement toward statistical process control in the early 1980s, and the movement has continued to gain momentum ever since.

The Statistics Behind Statistical Process Control

In the last chapter, we explored a few basic statistical concepts. One of these was the concept of a normal curve to represent a normal distribution. As you will recall, the concept of a normal distribution is that, in many situations, most values for any measurement tend to cluster around an average value (we called this the mean value). We also covered the concept of a standard deviation, and explained that the standard deviation represented a measure of the data's spread about the mean.

At this point, let's consider one other statistical concept, and that is what happens when a distribution is not normal. As explained in Chapter 9, most things in the world tend to follow the normal distribution. But not all do. Sometimes, groups of data are not normally distributed (for a lot of different reasons we won't go into here). That doesn't impede our application of statistics to

manufacturing processes, though. Even though data may not be normally distributed, we can still calculate averages for groups pulled from this distribution.

Why is being able to calculate averages important? There's a statistical concept called the central limit theorem that tells us that even if a group of data is not normally distributed, samples pulled from this group will have averages that are. If we don't work with the raw data in this non-normally distributed population, but instead only use data points calculated from the averages of groups pulled from the larger population, we can still use the concepts underlying the normal distribution to work with the averages. This is an important advantage for us, and it is an underlying concept that makes statistical process control possible. Statistical process control takes advantage of these concepts to place control of processes in the hands of the operators.

Taking off the Blindfolds

Remember the example at the beginning of this chapter about driving blindfolded? Suppose that instead of forcing someone manufacturing parts to manufacture them blindfolded, we instead allowed the operator to measure the parts as they were produced, and further allowed this operator to track the data. How could this be accomplished? Figure 10-1 illustrates a simple example. Suppose a machinist is manufacturing a shaft, and he plots the diameters of the shaft on a chart. If the values hover around a mean diameter, the machinist has some assurance that the process he is using (the lathe, the cutting tool, the type of metal being cut, etc.) is under control. Suppose something happens, though, that begins to affect the diameter of this hypothetical machined shaft. Let's say the cutting tool began to wear. The operator would see this wear if he plotted the shaft diameters, as the worn tool would produce shafts with larger and larger diameters. If the operator recognizes this upward trend (as shown in Figure 10-1), he could stop the process, search for the reasons it is starting to drift out of control, fix the problem, and prevent the creation of a nonconforming part.

What has been accomplished by the machinist described above? With this simple approach to manufacturing management, a simple approach to statistical process control, our machinist has accomplished a great deal. The organization employing the machinist moved squarely from a detection-oriented quality management philosophy to a prevention-oriented philosophy. The machinist has been empowered to recognize impending defects before they occur, and take steps to prevent their occurrence. This is a significant accomplishment, but something else significant also occurred. The need for someone to inspect the shaft after the machinist machined it has been eliminated. There is no need to inspect the item, because the machinist already knows it meets drawing requirements. So does anyone else who looks at the chart the machinist is maintaining.

Shaft Diameter
Machined Aluminum Shafts

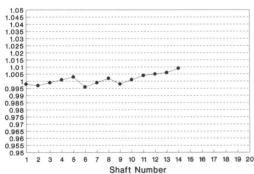

Figure 10-1. Machined Shaft Diameters. In this simple example, a machinist has measured shaft diameters and plotted them on a chart. Note that the diameters are starting to increase, which indicates that the process is drifting out of control.

There's more to statistical process control than this simple example shows (and we'll cover this shortly), but the management concepts and underlying purposes of statistical process control are no more complicated than our simple example illustrates. Defect rates are reduced significantly because the operator can prevent nonconformances, and the need for subsequent inspection of parameters under statistical process control disappears. Quality increases. Cost decreases. That's what it's all about.

Implementing Statistical Process Control

Implementing statistical process control is a six-step process that consists of selecting processes that are

candidates for statistical control, defining exactly what the process is, selecting the points within the process that are to be statistically controlled, training the operators, gathering data for those points to come under statistical control, and then preparing, maintaining, and using the charts to control the process. The process is illustrated in Figure 10-2. Each step is further explained below.

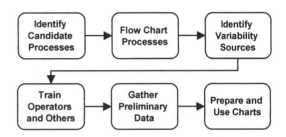

Figure 10-2. A Six-Step Process for Statistical Process Control Implementation. These steps assure critical parameters are identified and selected for SPC applications.

Selecting Processes for Statistical Control

One of the first steps in implementing statistical process control is realizing that the technique is not appropriate for all processes.

In our experience, processes ideally suited for statistical process control are repetitive (in the sense that they produce quantities of similar items), have a high inspection content, have higher than desired reject rates, and create items with dimensions or other characteristics that are fairly straightforward to measure. Processes with high inspection content are potential candidates because statistical process control greatly reduces or eliminates inspection. Processes producing parts with high reject rates are candidates because statistical process control, when properly implemented, frequently eliminates defects. Processes that are not producing rejects should also be considered for statistical control, but processes with high reject rates should be considered first.

Processes that produce items with dimensions or other characteristics that are fairly straightforward to measure are good statistical control candidates because straightforward measurement techniques simplify statistical process control training and acceptance.

Attributes Versus Variables Data

The last characteristic listed above for identifying statistical process control candidates (processes producing parts with dimensions or other characteristics that are fairly straightforward to measure) brings us to another issue. This issue concerns attributes and variables data. Variables data are related to measurements with quantifiable values (for example, shaft diameters are measured and recorded with specific values, as shown in Figure 1). Attributes data only reflect a yes or no decision, such as whether an item passed or failed a test. Attributes data are recorded in such terms as pass or fail, go or no go, yes or no, true or false, accept or reject, etc. There are no quantifiable values included with attributes data.

Statistical process control methodologies include different approaches for tracking attributes data versus variables data. These different types of statistical process control charts will be discussed in greater detail later in this chapter. For now, it's only important to recognize that the differences exist.

Defining the Process

Once a process has been selected for statistical control, the next step is to understand the process. This may sound like a trivial exercise, but our experience has proven that it is not. In many cases, no single individual understands the entire process. This occurs because processes evolve over time. Changes and improvements are incorporated, people are reassigned or leave the organization, and new people join the organization. The people assigned to maintain the process (or produce items using the process) frequently see only a small portion of the entire process. If there are factors upstream or downstream that affect hardware acceptability, individuals assigned to a process frequently do not have the visibility to recognize or understand these factors.

How does one go about defining the process? We have found that the best technique is one that has everyone with a stake in the process helping to create a detailed flow chart. When developing a process flow chart, we recommend including the operators, the area supervisor, the manufacturing engineer, the inspectors, the quality engineer, and the design engineer. One might be tempted to think that everyone of the above individuals (especially the manufacturing engineer and the operators) already understand the process and that creating a detailed flow chart is not necessary. Our experience refutes this belief. Every flow chart creation meeting we've observed resulted in every one of the participants being surprised at the extent and details of the process under evaluation (a process everyone previously thought they knew intimately).

Selecting SPC Implementation Points

The flow charting exercise described above is necessary to understand the process. This effort supports the selection of statistical process control implementation points. Selecting statistical process control implementation points involves understanding the part's critical parameters that are to be controlled (such as a shaft diameter or some other component characteristic), and then identifying the process parameters that influence this parameter.

In addition to implementing statistical process control on a component dimension, it may also be necessary to implement statistical process control on process parameters that influence the part (that's why it's important to understand the process). For example, dimensions on a cast part may be selected for statistical control, but an improved understanding of the casting process may show that casting temperatures strongly influence dimensional stability, and therefore, the casting temperature should also be selected for statistical control.

There are several techniques for selecting component and process parameters for statistical control. One of the most popular is the Ishikawa diagram (named after Dr. Kaoru Ishikawa, who was one of the leading Japanese quality consultants), or as it is more commonly known, the fishbone chart.

A sample fishbone chart for a metal forming process is shown in Figure 10-3. The fishbone chart is a graphical portrayal of the conditions that can produce a nonconforming item. One starts by drawing the fishbone "spine" (a horizontal line) with the part under consideration labeled at the right end. One then adds "ribs" to the chart to show various nonconformances. For each of these ribs, one then

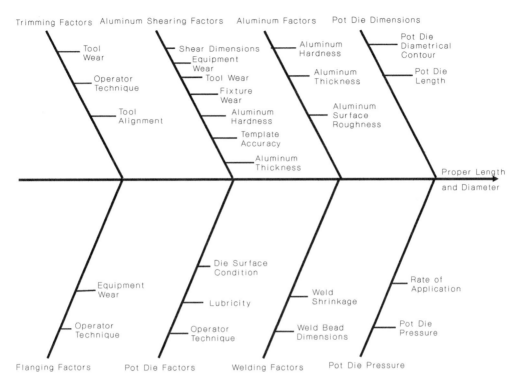

Figure 10-3. Sample Ishikawa Diagram. This diagram, commonly known as a fishbone chart, identifies causes of process variability. The fishbone chart shown here was prepared for a metal forming process.

adds branches that show the underlying conditions that can cause the nonconformance.

Fishbone charts are best created with the same team that created the process flow chart. This approach will help to create synergy in identifying potential nonconformance causes. Once these potential causes are identified, the statistical process control implementation point selection process can be performed in a more enlightened manner. Incidentally, fishbone charts are also useful for problem-solving and failure analysis, as discussed in Chapters 5 and 6.

Training the Operators

One of the most important elements of any statistical process control implementation program is training. The concepts involved in statistical process control are somewhat abstract and require an elementary understanding of the nature of statistical

distributions, as covered in the previous chapter. Training the operators who will be preparing the charts is most significant, as one of the underlying purposes of statistical process control is to place control and monitoring of the process in the operators' hands. If the operators do not understand the principles behind statistical process control, the process will more likely than not be viewed by the operators as just another chore, and not a tool to help them produce quality hardware.

What should a statistical process control training program consist of? We believe there should be three levels of training. The first should familiarize an organization's senior management with the purposes of statistical process control, its underlying principles, the approach to implement it, and how they can support the effort. The senior management training element can be fairly brief (one to two hours is typical, with follow-up implementation status reviews every month).

97

The second portion of the training program should be targeted for an organization's manufacturing and quality engineers. This training module should include the same subjects as the executive overview, but in much greater depth. It should also include detailed information on the makeup of the control charts (this will be covered in more detail shortly), problem-solving techniques, and problems likely to be encountered during the implementation effort (as well as recommended solutions for these problems). Most organizations implementing statistical process control invest in 20 to 40 hours of training for their manufacturing and quality engineers, with refresher training sessions as necessary.

The third portion of the statistical process control training program is presented to the operators. This portion should be tailored to the specific points for which statistical process control is being implemented (it should focus on the measurement techniques to be used for gathering data to be plotted on the control charts). It should also focus on the mathematics necessary for preparing the charts.

The mathematical operations associated with statistical process control might be viewed with some anxiety, but these are simple operations involving only addition and division (as we'll explain later). The training should present underlying statistical concepts to help operators understand the statistical process control approach. Operator training should emphasize how to plot data points on the charts, and most significantly, how to interpret the charts, how to spot trends that show a process is drifting out of control, and what to do when this occurs. Operators are frequently provided with 20 to 40 hours of initial training, with follow-up sessions of 1 to 4 hours on a monthly or quarterly basis.

Many companies ask if they should perform the training with in-house resources, or go outside for expert assistance. Whether to perform this training internally or externally (i.e., to rely on employees knowledgeable in statistical process control or to retain outside training organizations) has been a subject of considerable debate, particularly in medium to small companies. Most medium to small companies embarking upon a statistical process control implementation effort typically do not have the requisite internal expertise for this task.

As more and more companies implement statistical process control there is a larger pool of potential employees hiring into organizations with statistical process control experience, so this situation is changing. Small to medium companies typically have limited resources, and unfortunately, externally provided statistical process control training is not inexpensive. Our recommendation is to start with externally provided training, and then make a decision once the effort is underway whether to continue the training program with outside support or have employees conduct the training.

Gathering Data

Once the processes and the points in the processes to be brought under statistical control have been determined, the next step is to gather data. What data is gathered? Exactly the same information that will be collected once a process is monitored statistically, or brought under statistical process control. If one is attempting to statistically control the diameter of a machined shaft, then the diameters of a large number of shafts must be gathered. If one is attempting to statistically monitor the shutoff capacity of an aircraft external fuel tank, then one must collect shutoff capacity data prior to preparing a statistical process control chart.

The data described above is gathered to allow calculating mean values and upper and lower control limits for both variables charts and attributes charts (as will be explained immediately below); however, the methodologies for calculating the upper and lower limits for these two different types of charts are different. Although the temptation is great, one cannot simply begin charting data immediately. Data must first be accumulated to allow an accurate determination of what the process can provide.

Using Control Charts

Up to now, this chapter referred several times to variables data and attributes data, and the fact that different types of statistical process control charts are used for controlling each. Although there are many different types of control charts, the two most frequently used are the x_{bar}:r chart and the p chart. Readers interested in other statistical process control charts are invited to review the references listed at the end of this chapter, and in particular, the

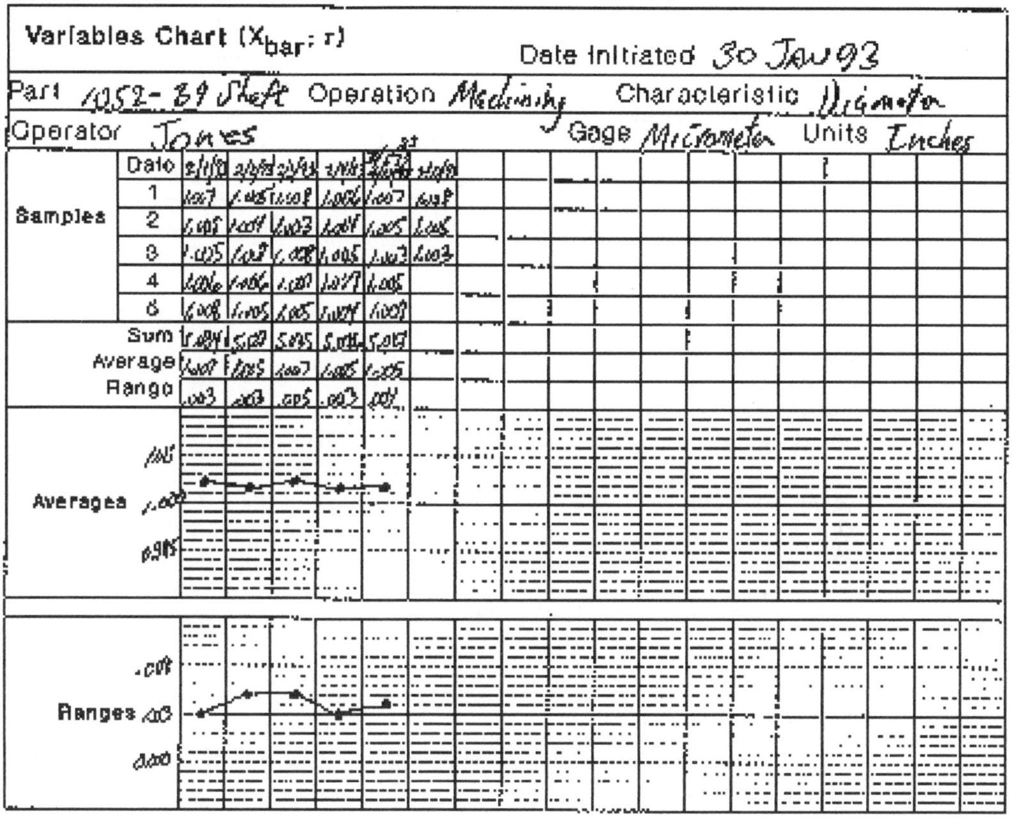

Figure 10-4. *x_{bar}:r Chart for a Machined Shaft. There are many elements of information on this chart, as explained in the accompanying text.*

DataMyte Handbook (which is one of the best books available on this subject). All of the other types of statistical process control charts are derivatives of the x_{bar}:r and p charts, so we'll focus on these two in our discussion. x_{bar}:r charts are used for controlling processes for which variables data is available, and p charts are used for controlling processes for which attributes data is available. Let's examine each in more detail.

Preparing an x_{bar}:r Chart

Figure 10-4 shows an example of an x_{bar}:r chart. Notice that the x_{bar}:r chart shown in Figure 10-4 contains many elements of information:

- The top of the chart includes data related to the process and the parameter being controlled (these data include such things as the date the

chart was initiated, the component being controlled, the parameter being monitored, the operator's name, and the machine being used to produce the component).

- Just below this data, we next see columns of individual values for the shaft diameters. There are five individual values in each column. At the bottom of each column, we see the total of the shaft diameters in that particular column, and then below that value, the average value of the shaft diameters in that column. This is the x_{bar} value that will be plotted below.

- In each column at the top of the chart, we see a value for the range of each column (in other words, the difference between the lowest value and the highest value). This is the "r" value that will also be plotted above.

- Moving up to the center of the chart, we see a strip of plotted data with boundaries identified as upper and lower control limits, and a mean value (how these are calculated will be discussed shortly). Within this strip, we see individual points plotted that correspond to the columns of data listed below. These individual points are the x_{bar} values that correspond to each column. We also see lines connecting the individual x_{bar} data points.

- Just below the columns of individual values, we see another strip of plotted data, also with boundaries identified as upper and lower control limits, and a mean value (the calculation of these values will also be discussed in just a bit). Within this strip, we similarly see individual points (and again, these correspond to the data in the columns below), but this time, the individual points represent the "r" values from each column. And again, we see lines connecting the individual "r" data points.

How are control charts such as the one shown in Figure 10-4 initiated? As explained above, prior to initiating the chart one must gather data. Most statistical process control practitioners recommend collecting at least 20 different subgroups of data. A subgroup of data is the data included in one column, which in turn typically includes five individual data points. This collection of 20 data subgroups would, therefore, represent 100 different individual data points. Once the 20 subgroups of data have been obtained, several calculations are performed to initiate a control chart.

The first calculation finds the average of the 100 individual data points (one accomplishes this by simply adding the 100 data points, and dividing the sum by 100). This is the grand average, and it becomes the value of the line in the center of the x_{bar} strip of data (as shown in Figure 10-4).

The next calculation determines the upper and lower control limits for the x_{bar} strip of data. This calculation is a little more involved than determining the grand average. There are two approaches to performing this calculation, but before these are discussed, one needs to understand what the upper and lower control limits represent. Upper and lower control limits represent three standard deviation

departures from the grand average (or the mean value of the population, as indicated by the sample of 100 individual data points). The first approach to determine the upper and lower control limits is to simply calculate the standard deviation of the sample of 100 data points (as described in the previous chapter), multiply this value by three, and then add the value to the grand average to determine the upper control limit, and subtract the value from the grand average to determine the lower control limit.

A second and slightly less complex approach is to use statistical process control tables to provide a value referred to as A_2. The A_2 value is a shortcut method for determining the plus-and-minus three sigma limits. One simply multiplies the A_2 value times the average range (we'll say more about this in a minute), and then adds and subtracts this value from the grand average to determine the upper and lower control limits. Tables of A_2 values are generally available in statistical process control texts or instructional materials. The A_2 values included in these tables make allowances for sample size to provide a realistic estimate of the standard deviation.

Regardless of the approach used for determining the x_{bar} upper and lower control limits, these values are simply drawn as dotted lines above and below the grand average line. This is shown in Figure 10-4.

The third calculation determines the value of the average range. This is calculated by determining the ranges for each of the 20 subgroups of data gathered earlier, summing these values, and then dividing the sum by 20. The resulting value is drawn as a line on the chart to represent r_{bar}, or the average range, as Figure 10-4 shows.

Finally, the fourth set of calculations involves determining the upper and lower control limits for the r_{bar} strip of data. As was the case earlier for x_{bar}, two approaches can be taken for determining the r_{bar} upper and lower control limits. Based on the 20 ranges obtained from the subgroup data (i.e., the data gathered prior to preparing the chart), one could simply calculate the standard deviation and determine plus-or-minus three standard deviation values to determine the upper and lower control limits. The other approach is to use D_3 and D_4 values, which are similar in purpose to the A_2 value described above. One simply refers to a statistical process control table to determine D_3 and D_4. These values are similarly based on subgroup size and provide a quick means of determining plus-and-minus three standard deviation control limits. We can simply multiply D_4 times r_{bar} to determine the upper control limit, and D_3 times r_{bar} to determine the lower control limit. Once these values are determined, lines are drawn on the r_{bar} strip to represent the r_{bar} upper and lower control limits, as shown in Figure 10-4.

Maintaining an x_{bar}:r Chart

Having gathered preliminary data and used it to define the grand average, the average range, and the upper and lower control limits for each, we can now begin plotting data and using the chart to control the process. Most of the hard work has been done. All the operator needs to do is record each measurement in the columns at the bottom of the chart.

Once five data points are collected, the operator finds the average and plots it in the x_{bar} data strip. The operator similarly determines the range and plots it in the r_{bar} data strip. The process is repeated for the next five measurements, and two new x_{bar} and r values are plotted. Lines are drawn between the two x_{bar} values. Lines are similarly drawn between the r values. The process is repeated for each new set of five measurements.

In addition to plotting the points as described above, it's also a good idea to train operators to note any changes in the process, tooling, operators, or other factors directly on the chart, with lines drawn to indicate where the change occurred. For example, if a cutting tool was changed in the lathe used to machine the shafts described in the example above, it makes sense to note this on the control chart. If an adverse trend develops (or perhaps, if the shaft diameter variability is reduced), it is of enormous benefit to be able to identify a change in the parameter being statistically controlled with a change. Noting changes on the control chart greatly facilitates making such determinations.

Using an x_{bar}:r Chart

Once operators have been trained, data has been collected, averages and upper and lower control

limits have been determined, and operators begin recording and plotting data, the statistical process control effort is ready to go to work for the operator. Many organizations err by believing they have implemented statistical process control once the above actions have been accomplished. Up to this point, though, all an organization has accomplished is to put itself in a position where it can begin to use statistical process control. Consider the name of this technology: statistical *process control*. Having accomplished the above, it's now time to let the operator control the process by using statistics.

How is the above accomplished? By teaching the operator (as well as the area supervisor, the quality engineer, and the manufacturing engineer) to use the data on the charts to prevent making bad hardware. Actually, this is where the fun begins, because most of the difficult work has already been accomplished, and from this point forward, it's simply a matter of plotting data and looking for trends in the data.

There are basically four types of trends to search for, and for the most part, these are found in the x_{bar} data. The first is the simplest, and that is no trend at all. If the subgroup averages and average ranges appear to be evenly distributed on either side of the nominal average (the center line, as shown in Figure 10-5) and the points remain inside the upper and lower control limits, then the process is in control. Ideally, the closer these points remain to the nominal, the more the process is in control. What this means is that the process is not likely to produce nonconforming hardware if operating conditions are stable (there are no changes).

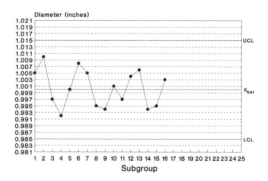

Figure 10-5. A Process in Control. This section of x_{bar} data shows an evenly distributed number of points above and below the x_{bar} line. This indicates the process is in control.

The second trend to watch for is either the subgroup averages or average ranges moving up or down, such that if the process continues it is likely an upper or lower control limit will be crossed. This condition is illustrated for a subgroup average in Figure 10-6.

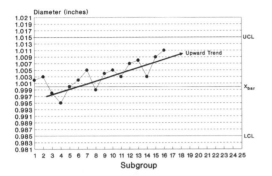

Figure 10-6. A Process Heading out of Control. If the subgroup averages start to trend toward the upper or lower control limits, it indicates the process is drifting out of control. This situation allows the operator to call for help before nonconforming hardware is produced.

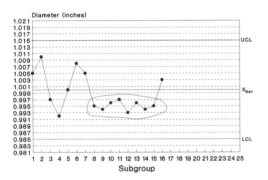

Figure 10-7. Another Indication of an out of Control Process. If seven or more sequential data points are above or below the x_{bar} line, something is causing a shift in the process nominal values. Corrective action should be taken before nonconforming hardware is produced.

Although the situation shown in Figure 10-6 is for x_{bar} data, it could also exist for the average range plot. Figure 10-6 shows that the process is starting to drift out of control, and if no changes are made, parts will soon begin to cross the upper control limit. This is where the preventive nature of statistical process provides the operator of a process under statistical control with a real advantage. The

operator can see things starting to go bad before nonconforming hardware is produced. As a result, the operator can take corrective action before any nonconforming hardware results. We'll cover how this can be accomplished shortly.

The third trend to search for is seven or more points in a row either above or below the center line. This is shown in Figure 10-7. It's highly unlikely this will occur without something in the process changing, and therefore, the presence of a run of points either above or below the centerline is an indication the process is drifting out of control. Again, when the operator sees this, it's time to call for support before nonconforming hardware is produced.

The fourth trend that operators should consider when reviewing their control charts for trends is a pattern that repeats. An example of such a condition is shown in Figure 10-8. This means that something is occurring in the process that causes the data to shift cyclically (instead of randomly), and if it is not fixed, nonconforming hardware could result.

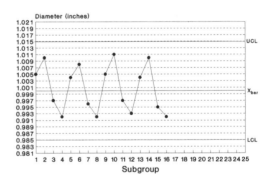

Figure 10-8. A Cyclical Indication of an out of Control Process. If a pattern repeats in the x_{bar} data, something is changing in the process. Again, corrective action should be taken before nonconforming hardware is produced.

Remember the example at the beginning of this chapter, with Arleigh Henderson, Ben Braddock, and the fuel tank shutoff capacity problem? Remember Braddock's comment about sensing that shutoff capacity failures were about to occur because the test values were trending up or down? Remember his comment:

You don't have to be a genius to see it coming.

That's what statistical process control is really about, as explained in the preceding paragraphs. It's a way of collecting data that shows if the process is starting to drift out of control, and doing something about it before nonconforming hardware is produced.

Calling for Help

Let's now return to our earlier question: How does the operator get help when statistical controls show the process is starting to drift toward producing nonconforming hardware? It depends on the nature of the process parameter or component characteristic under statistical control. If it's something as simple as a machined dimension starting to drift toward a control limit and the operator knows that cutting tool wear will induce the condition, the operator need only replace the cutting tool, observe that the process is back in control, and continue on. If the operator does not recognize the reasons for the process drift, then support from someone else is needed. This additional support can come from the area supervisor, the design engineer, the quality engineer, the manufacturing engineer, or perhaps others.

The key point here is that operators using statistical process control need to know that if they see their process drifting toward an out of control condition and they can't fix it, they need to call for help, and they need to do so quickly. This aspect of statistical process control is an important element of the training process, and it should become an important part of an organization's culture.

Two concepts are important in this philosophy of operators asking for help: Operators can't be afraid to ask for help, and those who provide support to the operators have to do so expeditiously. Engineers, quality engineers, manufacturing engineers, and shop floor supervisors have to understand that when it comes to keeping a process in control, the operator is the customer, and everyone else in a facility is there to provide support to the operator.

Once it has been determined that a process is starting to drift out of control, the most meaningful indicators of what might be causing the drift are the notations the operator has made indicating changes to the process, and the range data. The notations concerning changes and processes that start to drift

are almost self-explanatory (if a process drift coincides with a change, there is a good chance the change has induced the process drift).

The range data is more subtle. Process changes (even if not noted on the chart) will almost always induce a range change. If one sees a range change, it is a good indication that something has changed in the process, and the troubleshooting approach for a process that has started to drift is to back up on the range chart to find the time the change appeared, and then search for changes in that time period.

Preparing and Using p Charts

Let's consider again the aircraft fuel tank shutoff capacity problem discussed at the beginning of this chapter. One approach to implementing statistical process control in this area would be to chart shutoff capacity on an x_{bar}:r chart as discussed above. But suppose one is working with attributes characteristics, and the pass-fail criteria do not lend themselves to variables data and the x_{bar}:r chart approach described above. What can be done in this situation?

Suppose a company is involved in the manufacture of light bulbs, and the characteristic of concern is how many bulbs are broken when a box is packed at the end of the production line. Clearly, a broken light bulb is a characteristic that represents attributes data. It is either broken or it is not. One would not attempt to characterize how much the light bulb glass fractured.

There's another way to examine the problem, though, and that's by converting what appears to be attributes data to variables data. This can be accomplished by simply tracking the number of defective broken bulbs and determining the percent defective in a reasonable sample. Instead of simply noting if each bulb is broken, let's determine how many bulbs are broken in group of bulbs. If we convert this to a percentage, we now have variables data to work with.

One approach to tracking the light bulb percent defective data would be to simply use an x_{bar}:r chart, as described earlier. Another approach for statistically controlling the light bulb process, however, would be to use a p chart.

Here's how it works. The amount of nonconforming bulbs in the group is denoted by the fraction p. We can compute the percent defective, p, for several subgroups. One then determines the average percent defective by using the formula:

$p_{average} = (np_1 + np_2 + np_3 + np_4 + ... + np_k)/(n_1 + n_2 + n_3 + n_4 + ... + n_k)$

In the above formula, n represents the number of bulbs in each subgroup, and k represents the total number of subgroups. This number, $p_{average}$, is then drawn on the control chart as the average percent defective. It becomes the centerline, in the same manner as the grand average was plotted as the centerline for an x_{bar}:r chart. The concept is shown in Figure 10-9.

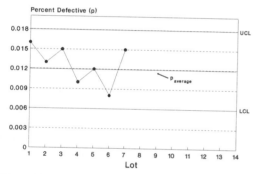

Figure 10-9. A p Chart for Defective Light Bulbs. This chart is used for attributes data, and it shows the percent defective.

The next step is to prepare the upper and lower control limits. The formula for the upper control limit is:

$$UCL = p_{average} + 3\sqrt{p_{average}(1 - p_{average})/n}$$

Similarly, the formula for the lower control limit is:

$$LCL = p_{average} - 3\sqrt{p_{average}(1 - p_{average})/n}$$

Once the upper and lower control limits have been determined, they are drawn as lines above and below the $p_{average}$ line on the control chart.

Having accomplished the above, the next steps are very similar to those performed for an x_{bar}:r chart. Instead of collecting five data points and finding the

average to plot a point (as was done for the x_{bar}:r chart), we simply need to calculate p for each subgroup and plot it on the control chart.

Again, once all of the above has been completed, one has only begun to implement statistical process control. The challenge now is to use the data to predict when one is starting to drift toward either the upper or lower control limits. The trends and evaluation approaches are similar to those described above for x_{bar}:r charts.

Implementation Challenges

Implementing statistical process control is a challenging process, and every organization we've observed going through this process is faced with a number of risks. These challenges include resistance to change, workers regarding the effort as simply another task, failure to maintain the charts, not using the charts as a preventive tool, and applying statistical process control in the wrong places. Each of these challenges, along with suggestions on how to meet them, are described below.

The immediate risk in implementing any change, including statistical process control, is that the change will be resisted. To be sure, statistical process control represents a major change to people used to working in an inspect and detect environment. Classical tools used for managing resistance to change are required here: training, coaching, helping those that might resist become a part of the implementation effort, and praising success stories as they occur (and they will occur).

We've found that seeking input from those doing the work (i.e., those who will be using statistical process control to control the quality of their work) when selecting statistical process control implementation points helps enormously. This makes shop floor workers participants in the change process, rather than simply being people directed to accept change. We also recommend publicizing each instance in which statistical process control reduced or eliminated nonconformances. We've used company newsletter articles, flyers, special letters, and award ceremonies in the work area for this purpose. The idea is that other people in the plant see the process work.

Another implementation risk is that preparing the charts degenerates into just another management-directed chore, without being of any real use to those who are preparing the charts. The best way to guard against this is to stay close to the effort to see where it is not working, and then applying the right resources to correct any training, application, or other shortfalls. Publicizing successes, as described above, also helps to prevent statistical process control from becoming another item on a task list for already busy people.

Sometimes in new statistical process control implementation efforts we find that the control charts are not kept current. If those responsible for charting the data are not committed to the program, this will most likely be the result. The actions described above, as well as those to be described below, will help to prevent this. Management can help by reviewing charts frequently, both to detect failures to keep the charts current and to reinforce the organization's commitment to the implementation effort. One word of caution is in order when charts that appear not to be current are encountered: The parts on the chart may not have been manufactured recently. If this is the case, the chart will show no recent data entries.

Perhaps the most common problem in initial statistical process control implementation efforts is not using the information on the charts to prevent defects. This often results when any of the above conditions are present. Any of the above recommended corrective actions will help to prevent this from occurring. If nonconforming hardware is created on a process being monitored by statistical process control, it helps to show workers where they might have detected an impending nonconformance before it occurred. It also helps if an organization's managers, quality engineers, and manufacturing engineers review control charts frequently to search for impending nonconformances. When potential out-of-control conditions are found, pointing them out to those doing the work will support the implementation effort.

The last common problem we've observed involves inappropriate statistical process control implementation points. Often, in an organization's zeal to implement statistical process control, too many charts or otherwise inappropriately targeted implementation points result. If a process being statistically monitored still yields an unacceptably high reject rate, it's probably because the parameters being tracked on the chart are inappropriate. If workers are spending large amounts of time maintaining statistical process control charts, there are probably too many charts.

Summary

Statistical process control is a tool that emerged in America and migrated to Japan. It was ignored in America for many years while it helped Japan become a world quality leader. America re-embraced statistical process control in the last decade to help in the quest for continuous improvement.

One of statistical process control's key advantages is that it places the responsibility for quality squarely in the hands of the operator. Another key advantage is that it allows operators to determine if a process is drifting out of control before defective hardware is made, and in so doing, allows the prevention (rather than detection) of defects. The bottom line is that statistical process control allows the people doing the work to know they are producing conforming product, and to take preventive actions as processes show signs of drifting out of control.

References

DataMyte Handbook: A Practical Guide to Computerized Data Collection for Statistical Process Control, Edition 4, DataMyte Corporation, 1989.

Statistical Quality Control Handbook, AT&T, Delmar Printing Company, 1984.

Well Made in America, Peter C. Reid, McGraw-Hill Publishing Company, 1990.

Statistical Process Control Implementation Plan, Sargent-Fletcher Company, 1990.

Statistical Quality Design and Control, Richard E. DeVor, Tsong-how Chang, and John W. Sutherland, Macmillan Publishing Company, 1992.

Chapter 11

ANOVA, Taguchi, and Other Design of Experiments Techniques

Finding needles in haystacks...

Peter O'Brien stared in disbelief at the aerial refueling system on the test stand. As the project engineer assigned to upgrade the system, he was dumbfounded at the extent of the problems he faced. The system was not a new design. In fact, it had been in production for several years, and it was based on technology that had been used in previous systems for nearly three decades. Over the last several months, O'Brien had resolved many of the system's longstanding technical deficiencies, but one remained and it was stubborn. That was the explosively actuated hose guillotine.

The aerial refueling probe and drogue system operated by extending a 50-foot hose behind the tanker aircraft. Refueling occurred when the receiving aircraft flew into position behind the tanker and plugged a probe into the paradrogue at the end of the refueling hose. When the connection was complete, the tanker aircraft could begin pumping fuel to the receiver aircraft.

O'Brien's immediate problem was the aerial refueling system guillotine (an exploded drawing of the guillotine is shown in Figure 11-1). The guillotine was located at the end of the aerial refueling system pod. The guillotine surrounded the hose as it extended from the aerial refueling system, and its sole purpose was to clamp and jettison the hose if the tanker aircraft pilot commanded it to do so during an emergency. The guillotine was one of the aerial refueling system's key safety features. If the hose could not be retracted into the aerial refueling system due to a system failure, the guillotine's job was to separate it from the pod so that the tanker pilot could make a safe carrier landing. The guillotine was supposed to clamp the

hose so it would not leak, and then cut it in half to allow the trailing portion to fall away.

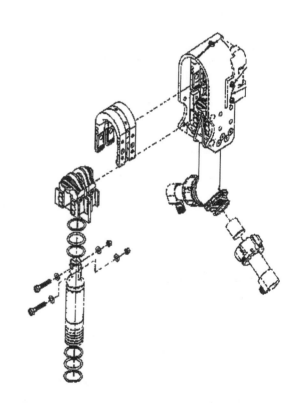

Figure 11-1. Aerial Refueling System Guillotine. The guillotine surrounds the hose at the end of the aerial refueling system. Its job is to cut and clamp the hose during an emergency hose jettison.

O'Brien looked at the aerial refueling system on the test stand with dismay. He had just completed another guillotine test, and the results were bad. The guillotine fired, but the hose had not separated from the store, and it was leaking fuel. O'Brien shook his head. In previous tests, the guillotine had clamped adequately, but it had failed to cut the hose cleanly. Changes were made to the cartridges in the guillotine to increase the explosive output. This was supposed to fix the hose cutting failures, but then the hoses wouldn't clamp properly and they leaked fuel. O'Brien next modified the clamp geometry, which fixed the clamping problem on two tests, but on the third the guillotine again failed to cut the hose cleanly. O'Brien next incorporated subtle design modifications to address both the cutting and the clamping problems, but this resulted in yet another failure.

The aerial refueling system program manager, Dwayne Dunford, walked up to O'Brien at the test stand and sensed his dismay. "Not a good day, huh, Pete?" Dunford asked.

"I can't seem to get to a reliable guillotine configuration, Dwayne," O'Brien answered. "If I do something to get the clamping feature to work, it fails to cut reliably. If I do something to fix the hose cutting feature, then it starts to leak again. My problem is that I've probably got four or five things I can change in the design that affect both the cutting and the clamping features, but I can't seem to get the right combination to get the design right. There's a combination of four or five things that have to work together so that the guillotine will both cut and clamp reliably. If I fix one, I hurt another. If I then fix that one, something else goes wrong. It's a real bear."

"So your problem is you have several design inputs that are affecting one or two outputs, and you don't know how to optimize the design for the two outputs?" Dunford asked.

"Exactly," O'Brien answered.

"How are you approaching the testing on this device?" Dunford asked.

"I'm trying to use a very disciplined approach," O'Brien said. "I'm only changing one variable at a time in the guillotine design, because if I change more than one, I won't know which is having an effect."

"Makes sense to me," Dunford said, "but with four or five variables, won't you have to do a lot of tests?"

"Yes," O'Brien answered, "and therein lies at least part of the problem. I'm only doing one or two tests with each new combination of variables, and I don't think I'm getting a statistically meaningful result. Maybe it works one time, but I don't have any confidence the results will be repeatable. I'm not even sure when I get a success it means anything. We've been having random failures, maybe we're also having random successes."

"Is there any other test approach we could use?" Dunford asked. "We can't be the first company to struggle with this kind of problem."

The Nature of Experimental Design

Experimental design, or design of experiments, is a complex subject. Understanding this complex subject is critical in the quest for quality improvement. Whether we seek to improve a design or a process, we need good data upon which to make decisions.

When faced with opportunities to improve a design or a process, we frequently make tentative conclusions about the parameters that affect how well the product or process performs. In experimental design, we seek to test our assumptions about these parameters. If our assumptions prove to be correct, then we know that the parameter we suspected of making a difference in product or process performance is truly one that can influence performance. We would then control the parameters we found to make a difference such that product or process performance met our requirements.

We have observed that in many organizations the process for testing assumptions about parameter influence is less than rigorous. Our objective in this chapter is to add legitimacy to the experimental design process, and to help our readers perform tests that provide meaningful results. These results, in order to be meaningful, frequently must be rooted in statistical analysis. We've already had an introduction to statistical analysis and statistical process control in earlier chapters. At this point, we are going to explore how we can use concepts associated with the normal curve to help us make informed product and process improvement decisions.

The Null Hypothesis

There is one other concept we must first examine before we explore several popular experimental design statistical analysis techniques. That concept involves the null hypothesis.

When we suspect that a parameter makes a difference in how a product or process performs, we are hypothesizing that the parameter has an influence. For experimental design purposes, we need to formalize the nature of our hypothesizing.

The hypothesis formalization process occurs by assuming that a certain parameter makes a difference. We then hypothesize that a change in the parameter makes *no* difference in product or process performance (the assumption that the parameter makes no difference is why the hypothesis is called the null hypothesis). We then select an appropriate statistical analysis technique (we'll say more on this later) to evaluate the null hypothesis. We then perform the test and collect the data. Using the selected statistical analysis technique, we then assess the data to either accept or reject the null hypothesis. Figure 11-2 shows the concept.

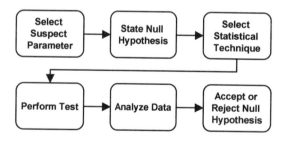

Figure 11-2. The Experimental Design Approach. The process consists of six steps, as explained above.

The null hypothesis allows us to make informed decisions about the experiment's results. If we think a parameter makes a difference in product or process performance, we hypothesize that it does not. If the statistical analysis technique we selected prior to

performing the test shows, statistically, that the probability the parameter does not make a difference is low, we reject the null hypothesis and conclude that the parameter does make a difference. If the statistical analysis technique shows that the probability the parameter does not make a difference is high, we accept the null hypothesis and conclude that the parameter makes no difference.

There is an important caveat we must add at this point, however. Strictly speaking, we are on safe ground if the data show that we can accept the null hypothesis when the data show that the parameter does not make a difference. Things are less absolute when the data show that we should reject the null hypothesis (i.e., the suspect parameter makes a difference). The reason for this is that there may be another unknown variable that is making the difference. When we reject the null hypothesis and conclude that the parameter we tested makes a difference, the risk we take is that there is some other parameter that is making the difference. There's a risk that our tested parameter really is not the parameter influencing performance.

In practice, however, the risk in rejecting the null hypothesis and concluding that the suspect parameter makes a difference is accepted in most experimental scenarios. For completeness in understanding the experimental design approach, however, the reader needs to understand that such risk exists.

Selecting an Appropriate Statistical Technique

While there are large numbers of statistical analysis techniques, for our purposes we can concentrate on four:

- The Z-test
- The T-test
- The F-test
- Taguchi Analysis

We have found the above four techniques to be the most commonly used, and we have found that they allow managers and engineers seeking to improve products and processes to make informed decisions in nearly all product development and manufacturing situations.

The Z-Test

Suppose we have a situation in which we have a product that has performed in a certain manner, and we wish to determine if its performance is representative of the parent population. Stated differently, we might want to know if the item came from the parent population, or if its performance is sufficiently different to conclude that it did not come from the parent population.

To make this situation more understandable, let us assume that we are in the business of making bonded composite metal-and-fiberglass helicopter main rotor blades. As part of the blade manufacturing process, each blade is x-rayed to assess the bond line width (which is critical to blade strength). After manufacturing these blades for many years, we know that the mean blade bond line width is 0.440 inch, and the standard deviation is 0.008 inch. Let us further assume that we have a blade returned from service, and in examining it, we find that the blade has a bond line thickness of only 0.375 inch.

Our concern on finding such a narrow bond line is this: Is this blade from the same parent population as all of the other blades, or is there enough of a difference in this blade's bond line width to allow us to conclude that the blade did not come from the parent population?

The Z-test is the most appropriate technique to use in this case, because we have a data point, we know the mean and the standard deviation of the parent population, and we wish to determine if the blade came from the parent population. The Z-test is a simple approach, and it relies exclusively on the normal curve approach developed earlier in this book.

In our earlier discussion, we learned that the normal curve had two important parameters that completely defined it (the mean and the standard deviation). We further learned that if we knew the mean and the standard deviation, we completely defined the normal curve for a particular distribution. Finally, we learned that the number of standard deviations away from the mean corresponded to the probability of that data point being realized.

The Z-test uses the above properties to assess the likelihood that a particular point is from a parent population. Here's how it works:

1. The first step is to select the suspect parameter. In this case it is blade bond line width.

2. The second step is to state the null hypothesis. In this case, the null hypothesis is that there is no difference between our suspect blade and all others. In other words, it came from the same parent population.

3. The third step is to select the appropriate statistical analysis technique. In this case, we select the Z-test. There's a sub-step to this test, and that is to select a confidence level. Typically in such development or industrial applications, a 95 percent or 99 percent confidence level is used. The confidence level simply represents the degree of assurance we are correct. If we select a 95 percent confidence level, then we will be 95 percent certain our conclusion is correct.

4. The fourth step is to perform the test, which in our case has already been accomplished. The result is the single data point that shows a 0.375 inch bond line width.

5. The fifth step is to analyze the data. In this case, we need to convert our single data point to a Z-value. We do so using the following formula:

$$Z = (x - \upsilon_x)/\sigma_x$$
where:
$Z =$ the Z-value
$X =$ the test statistic (in this case, 0.375 inch)
$\upsilon_x =$ the parent population mean
$\sigma_x =$ the parent population standard deviation

Computing the Z-value for our situation, we have:

$$Z = (0.375 - 0.440)/(0.008) = -8.125$$

Once we have the Z-value, we now have to perform the Z-test. There are two ways to do this. We can either use a standard Z-table (available in any statistical and many mathematics texts), or we can use the convenient Excel spreadsheet NORMSDIST(Z-value) function.[2] Using the latter approach, we state in an Excel worksheet:

$$= \text{NORMSDIST}(-8.125)$$

and see that the value returned is 2.22045E–16, which is a very small number.

6. The sixth and final step is to determine whether we should accept or reject the null hypothesis. In our case, the null hypothesis was that there was no difference between the 0.375-inch bond line width main rotor blade and the parent population, which we stated at a 95 percent confidence level. Since we used the 95 percent confidence level, we know that if the Z-value is 5 percent or less, we can reject the null hypothesis. In this case, the Z-value is considerably less than 5 percent (it is actually 0.0000000000000022 percent), so we reject the null hypothesis and conclude that the 0.375-inch blade is not from the parent population.

Having completed the above, we should point out a requirement for using the Z-test: One must know the mean and standard deviation of the parent population.

The T-Test

Suppose we encounter a situation in which we do not know the parent population's standard deviation, but we have a test sample and we would like to determine if our test sample is from the parent population. In such situations, one would normally have a small set of data points in the test sample. We can use the test sample to generate a standard deviation, and then use this information to generate a T-value in a manner similar to that used for generating the Z-value. The formula for computing the T-value is:

$$T = (\upsilon_s - \upsilon_x)/(\sigma_s/(n-1)^{1/2})$$

where:

$T =$ the T value
$\upsilon_s =$ the sample mean
$\upsilon_x =$ the parent population mean
$\sigma_s =$ the sample standard deviation
$n =$ the sample size

The approach and the six steps used with the T-test are identical to those identified for the Z-test. In a similar manner, once the T-value has been calculated, one can use either a standard statistical table or the Excel spreadsheet TDIST function.[3]

The F-Test and ANOVA

The next common experimental design technique we will address is analysis of variance (or ANOVA, for short). This technique is used in situations in which we wish to consider if differences in average performance between two or more groups is due to randomness alone, or randomness plus some special cause.

Remember the discussion in Chapter 9 that addressed the nature of variability and put forth the notion that all parameters have an inherent measure of variability. This inherent variability is induced by normal statistical randomness. Considering the above, one can understand that in all experiments there will be a measure of variability associated with each experimental variable.

Let's consider an example similar to the one at the beginning of Chapter 9, in which a room full of people and their weights were considered. For the people in that room, we calculated the average weight and introduced the normal distribution concept with a mean and a standard deviation about the mean. We saw in Chapter 9 that the average weight for the 30 people was 167.53 pounds, and we can calculate the standard deviation using the formula provided in Chapter 9 to find that it is 24.9 pounds.

Now, let us divide the room evenly in half to create two groups of 15 people each, and consider the distribution of weights for the people in each group.

---

[2] One can use the NORMSDIST function to create a table of Z-values. To find out how, or to obtain such a table, please contact our website at www.bhusa.com.

[3] One can similarly use the TDIST function to create a table of T-values. To find out how, or to obtain such a table, please contact our website at www.bhusa.com.

This data is shown in Figure 11-3. The first group's average weight is 172.26 pounds, and the second group's average weight is 162.80 pounds. The averages are pretty close, and due to the manner in which the group was arbitrarily divided, it seems logical to assume the difference in average weight doesn't really mean anything. We also show the standard deviations for each group, which were calculated using the formula in Chapter 9.

Consider the statement in the last paragraph. The difference in average weight does not really mean anything. It seems intuitively correct to make the above statement. In the context of everything discussed so far, what's really being inferred here is that there will be a measure of variability associated with the weights of everyone in the room, and therefore, with the average weights of any two or more groups taken from the original larger group of 30 people. The difference in average weight is probably due to normal statistical randomness. It is not likely that the average of any two groups would be exactly the same, so assuming that a small difference between the average weights of two groups is due to randomness alone is probably an acceptable conclusion.

Let us now consider the average weights for the same two groups of people, except this time, a set of factors were induced that might have influenced the two groups' average weights. Suppose the first group did no physical exercise for six months, ate all of their meals at fast food restaurants, and drank several beers each evening while watching television. The second group stayed with a different regimen for six months. This group ran two miles every day, drank no alcoholic beverages, and ate only high-fiber, low-sodium, low-cholesterol meals. The first and second groups' weights after their six month regimens are also shown in Figure 11-3.

As was the case previously, there's a difference in the average weights of the two groups (except that now it's much larger: 195.73 pounds versus 149.93 pounds). For these two groups, it seems logical to assume the differences in diet and physical exercise probably played a role. But what about the normal statistical randomness mentioned earlier? Might this also play a role in determining the difference between the two groups' average weights?

The answer to the above question is yes. Normal statistical randomness will always be present.

When the first two groups of 15 people were considered (when the room was arbitrarily split into two groups), it seemed logical to infer that the difference in average weight was due solely to this statistical randomness.

When the second two groups were considered (the fast-food couch potatoes and the physical fitness enthusiasts), it seemed likely to assume the differences in lifestyle contributed to the differences in average weight. It is important to recognize, though, that even for the second group of people we can assume that some statistical randomness will be present, and that this contributed to the difference between the two groups' average weights as well.

Prior to Regimen		After Regimen	
Group 1	Group 2	Group 1	Group 2
142	142	181	136
228	171	235	161
176	138	194	134
193	145	208	139
166	172	191	155
152	133	187	128
169	180	195	162
193	165	210	154
158	169	193	153
178	196	199	166
156	134	186	128
148	163	179	147
131	158	158	150
169	210	201	175
225	166	240	161
Mean: 172.26	162.80	197.13	149.93
St Dev: 27.01	21.50	19.98	13.82
Grand Average : 167.53			

Figure 11-3. Data For a Group of 30 People. The weights of 30 people, along with the means and standard deviations for each group of 15, are shown above.

When faced with differences in averages between two or more groups, a question naturally arises: Is the difference in average performance due to statistical randomness alone, or are there causal factors at work (in addition to the statistical randomness) that induced a change? This is the challenge that resulted in the statistical technique called analysis of variance, or ANOVA.

ANOVA Explained

ANOVA is an analytical technique that determines if differences in average performance are due to randomness alone or to specific causes. To do this, one needs a methodology for mathematically comparing the differences between the groups being considered. This methodology should consider both the spread of the data in each group, as well as the average for each group. If the spread of the data for one group is about the same as the spread of the data for another group, and if the average for the first group is about the same as the average for the second group, then it seems logical to conclude that the two groups are really not significantly different, and any differences that exist are due to statistical randomness.

As these differences increase (i.e., the differences in either the groups' averages or the spread of the data in each group), at some point it will be logical to assume that the differences between each group are due to one or more causal agents. Stated differently, at some point the differences between groups will be due to more than just statistical randomness. The challenge now becomes one of determining the appropriate mathematical methodology for making this determination.

Fortunately for us, the methodology for making this determination has already been developed by Sir Ronald Fisher, a British statistician. In the 1930s, Fisher was involved with agricultural experiments, and he needed to determine if differences in average crop yields were due to normal statistical randomness, or if changes induced by the experimenters to improve yield were making a difference. Fisher developed a technique for this purpose that became the framework for all modern variance analysis. The technique consists of six steps. This process is somewhat confusing, so we will continue our discussion using the two groups of people and their weights from our earlier example to clarify the approach.

- The groups to be evaluated are determined. These might be the two different groups of people described earlier, the crop yields from

different fields, or any other groups whose average performance is to be compared. In our case, we are going to examine the differences in average weight for the two groups of people described earlier after they followed their six month regimens.

- The parameter of interest is identified and quantified. Again, the parameter might be weight, crop yield, or any other parameter one wishes to compare. In our example, weight is the parameter of interest.

- The average value of the parameter for each group is determined. We found the average weight for the first group to be 197.13 pounds, and the average weight for the second group to be 149.93 pounds, as Figure 3 shows.

- The average for all of the values in all of the groups is determined (this value is frequently referred to as the grand average). In our case, this is the average weight of all 30 people after the six month regimens. As Figure 3 shows, the average weight for all 30 people is 167.53 pounds.

- The column sum of the squares is determined. This is done by finding the difference between the average for each group and the grand average, squaring the result, multiplying the result by the number of data points in each group, and then adding the values (this seems complex, but it will become simpler in a moment). Mathematically, the column sum of squares represents the variability of the group averages from the grand average, and it is denoted by SS_c. In effect, it creates a variability comparison between the groups' averages and the overall average. In our case, the difference between the first group's average and the grand average is $167.53 - 197.13$, or -29.60 pounds, and the difference between the second group's average and the grand average is $167.53 - 149.93$, or 17.60 pounds. These values are squared (to give 876.16 and 309.76), multiplied by the number of data points in each group (which is 15) and then added to give 17788.80. The calculations are summarized in Figure 11-4.

- The sum of the squares for the error is determined. Error, in the statistical and ANOVA context, does not represent mistakes in the calculations or imply the analyst is doing something wrong. Error is simply a statistical term that means the variability due to randomness alone. The error sum of squares is determined by finding the difference between each data point and its group average (for every data point in all of the groups being analyzed), squaring these values, and then adding them all together. The error sum of squares represents the total variability of the individual values from their respective group averages, and it, in effect, creates a variability index for the randomness due to normal statistical variation. This value is typically denoted by SS_e. In our case, each of the data points in the first group is subtracted from 197.13 pounds and the result is squared. We do the same in the second group, and then we add all of the values together. The results are summarized in Figure 11-5.

- The total sum of squares is determined. This is done by finding the difference between each value and the grand average, squaring the result, and adding all of the squared differences together. The total sum of squares represents the total variability of the individual values from the grand average. It creates a variability index for all of the variability in the groups being analyzed. The total sum of squares is usually denoted by SS_T. This is shown in Figure 11-6.

At this point, three sums of squares have been calculated: SS_T, SS_c, and SS_e. There's an important property of these sums of squares we need to address, and that is that the total sum of the squares is equal to the sum of the squares for the columns and the sums of the squares for the error:

$$SS_T = SS_c + SS_e$$

Readers will note the sums of squares we calculate do not exactly add up, which is due to rounding error. Recall that the sums of the squares are measures of variability. The above means the total variability is equal to variability due to special causes unique to the different groups analyzed (there may or may not be any), and the variability that exists due to normal statistical randomness.

One might ask where we're going with the above, and in particular, why everything is being squared.

After Regimen	Column Sum of Squares Calculations
Group 1	**Group 1**
181	$167.53 - 197.13 = -29.60$
235	$(-29.60)(-29.60) = 876.16$
194	$(15)(876.16) = 13,142.40$
208	
191	
187	
195	
210	
193	
199	
186	
179	
158	
201	
240	
Average: 197.13	
Group 2	**Group 2**
136	$167.53 - 149.93 = 17.60$
161	$(17.60)(17.60) = 309.76$
134	$15 * 309.76 = 4,646.40$
139	
155	
128	
162	
154	
153	
166	
128	
147	
150	
175	
161	
Average: 149.93	$SS_c = 13,142.40 + 4,646.40 = 17,788.80$

Figure 11-4. Computing Column Sum of Squares. Using the methodology explained in the text, the column sum of squares is computed as shown above.

Let us not lose sight of the fact that all of the above is being prepared to allow for determining if the differences between groups' averages are due to randomness alone, or randomness and special causes. The approach described above is finding differences between averages and actual values, and because of this, these differences represent a consistent standard of comparison. Why is everything being squared? The values are squared to both eliminate the effects of any negative numbers that might result when performing the subtractions, and to magnify the differences.

After Regimen	Data Point - Group Average	Squared Differences
Group 1	Group 1	Group 1
181	-16.13	260.18
235	37.87	1434.14
194	-3.13	9.80
208	10.87	118.16
191	-6.13	37.58
187	-10.13	102.62
195	-2.13	4.54
210	12.87	165.64
193	-4.13	17.06
199	1.87	3.50
186	-11.13	123.88
179	-18.13	328.70
158	-39.13	1,531.16
201	3.87	14.98
240	42.87	1,837.84
Average: 197.13		
Group 2	Group 2	Group 2
136	-13.93	194.04
161	11.07	122.54
134	-15.93	253.76
139	-10.93	119.46
155	5.07	25.70
128	-21.93	480.92
162	12.07	145.68
154	4.07	16.56
153	3.07	9.42
166	16.07	258.24
128	-21.93	480.92
147	-2.93	8.58
150	0.07	0.00
175	25.07	628.50
161	11.07	122.54
Average: 149.93		
		$SS_e = 8,856.67$

Figure 11-5. Computing Error Sum of Squares. SS_e is calculated using the methodology explained in the text, as shown above

Let us now turn to the degrees of freedom concept. Degrees of freedom is a somewhat misleading expression, as it merely describes the number of independent comparisons that can be made among variables and test specimens. This concept of *independent* comparisons is important, as it allows understanding an otherwise difficult to understand concept.

The simplest number of degrees of freedom is that associated with the total number of tests being performed or specimens being analyzed. This represents the total degrees of freedom, and it is

always equal to the total number of specimens minus one. In our weight example, there are 29 total degrees of freedom (or 30 minus 1). That's because the number of independent comparisons is represented by a comparison of each specimen to the next. Suppose we have eight test specimens, as shown in Figure 11-7. Only seven independent comparisons can be made. If any more comparisons are made, they are no longer independent, because the comparisons will be between two specimens previously compared to a third (and therefore, this comparison would not be independent).

After Regimen	Data Point - Grand Average	Squared Differences
181	13.47	181.44
235	67.47	4,552.20
194	26.47	700.66
208	40.47	1,637.82
191	23.47	550.84
187	19.47	379.08
195	27.47	754.60
210	42.47	1,803.70
193	25.47	648.72
199	31.47	990.36
186	18.47	341.14
179	11.47	131.56
158	-9.53	90.82
201	33.47	1,120.24
240	72.47	5,251.90
136	-31.53	994.14
161	-6.53	42.64
134	-33.53	1,124.26
139	-28.53	813.96
155	-12.53	157.00
128	-39.53	1,562.62
162	-5.53	30.58
154	-13.53	183.06
153	-14.53	211.12
166	-1.53	2.34
128	-39.53	1,562.62
147	-20.53	421.48
150	-17.53	307.30
175	7.47	55.80
161	-6.53	42.64
	$SS_T=$	26,646.67

Figure 11-6. Computing Total Sum of Squares. SS_T is calculated using the methodology explained in the text, as shown above.

Comparisons can also be made between the levels for each factor. In our weight example, there are two factors: the group that exercised, and the group that did not. One independent comparison can be made between these groups, so the factor degrees of freedom is equal to one. Incidentally, this is also referred to as the column degrees of freedom, since the columns are assigned to the factors when performing analysis of variance.

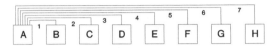

Figure 11-7. The Degrees of Freedom Concept. If eight independent variables are present, only seven independent comparisons can be made. Any additional comparisons are no longer independent, as they are influenced by the previous comparisons.

At this point, one more statistical rule should be incorporated, and that's the concept that the degrees of freedom for the error and the factors always equals the total degrees of freedom. This is similar to the rule described for the sums of squares, and it can be presented as follows:

$$df_T = df_F + df_e$$

where:

df_T = total degrees of freedom
df_F = factor degrees of freedom
df_e = error degrees of freedom

In our example ANOVA on the two groups, there are 29 total degrees of freedom. As we explained earlier, there is one factor degree of freedom (the number of independent comparisons that can be made between the two groups). Therefore, the error degrees of freedom has to be 28 (or 29 minus 1).

Having developed the above discussion, the analysis of variance can now be continued. Two more rules are necessary, which are stated below:

- Variance is defined as the sum of squares for a particular parameter divided by its corresponding degrees of freedom.

- The F-ratio is defined as the column (or factor) variance divided by the error variance. In our example, the F-ratio is 17,788.80 divided by 316.31, or 56.24.

Figure 11-8 summarizes all of our calculations.

	Sum of Squares	Degrees of Freedom	Variance	F-Ratio
Total:	26,646.67	29	918.85	
Columns:	17,788.80	1	17,788.80	56.24
Error:	8,856.67	28	316.31	

Figure 11-8. ANOVA Summary. The table summarizes the analysis of variance for the two groups' average weights.

The last topic we need to consider in our analysis of variance is the F-ratio. The F-ratio (named after Fisher) is used as a basis of comparison to determine if the difference between the groups being analyzed is due to randomness alone or to special causes. Here's how it works:

- One computes the sums of the squares and the variances to calculate an F-ratio as defined above. One could also use the Excel spreadsheet FINV function for this test.

- Once the F-ratio is determined, one compares this to a table of F-ratios or a F-ratio determined using the Excel spreadsheet FDIST function.[4] The F-ratio tables can ordinarily be found in any statistics text, and the Excel FDIST function, of course, is an integral part of the Excel software package.

- If the calculated F-ratio is larger than the value in the table or the value provided by the FDIST function, then the difference between the groups being analyzed is statistically significant. If the calculated F-ratio is smaller than the value in the table or the value provided by the FDIST function, then the difference between the groups is due to randomness alone. One selects the appropriate F-ratio from the statistics tables based on the degrees of freedom for the error and the degrees of freedom for the factors. The Excel FDIST function uses these values as inputs in determining the calculated F value.

Referring to the F-ratio table for our situation (28 degrees of error freedom and one degree of factor

---

[4] Alternatively, one could use the Excel spreadsheet FINV and FDIST functions to first calculate an F-ratio and then determine the probability that the two groups are from the same parent population. Please visit our website at www.bhusa.com for an Excel template to perform this analysis.

freedom), we find the tabular value to be 7.64. Our calculated F-ratio is 56.24, which is larger than 7.64. That means that the difference between the two groups we analyzed is not due to randomness alone.

What does the above mean to us, and how is it used in the real world? After all, we could have guessed that the difference in average weights between the two groups was not due to randomness alone, but was instead due to a special cause (the two vastly different lifestyles the groups followed in the preceding six months). Our null hypothesis would have been that there was no difference in the two groups' average weight; the ANOVA causes us to reject the null hypothesis and conclude that the difference in the two groups' average weight is due to randomness plus a special cause. Since the lifestyle of the two groups was the parameter we modified in performing the experiment, we conclude that the two groups' different lifestyles constitute the special factor influencing the average weight.

In the real world, we are frequently faced with situations where there appears to be a difference in average performance between groups, but we don't have the luxury of knowing beforehand if the differences are significant (and if they are, what the special causes are). When faced with this situation, one can perform analysis of variance to determine if differences in average performance are due to randomness alone, or if the differences are due to randomness combined with special causes. The approach in such a situation is to perform an analysis of variance, and if the differences are statistically significant, to then attempt to isolate them.

Dr. Taguchi's Quality Engineering Contributions

No experimental design discussion would be complete without a review of Dr. Genichi Taguchi's contributions to quality engineering and experimental design. The story of this remarkable man and his work, like many quality management building blocks, goes back to World War II and the effort to rebuild post-war Japan. As you will recall from our previous discussions, many events were under way in post-war Japan to set the stage for a remarkable industrial revitalization, with superior quality emerging as a cornerstone of Japan's business development strategy.

In addition to MacArthur bringing Deming to Japan (and Japan's acceptance of Deming's precepts on quality), Japan implemented a number of quality improvement initiatives. One of the very first of the Japanese quality improvement initiatives involved an effort to upgrade the country's telecommunications system. Taguchi headed Japan's communications research and development activities, and as such, he became the leader of Japan's telephone communications upgrade effort.

Taguchi's early telecommunications work formed the foundation for much of his quality engineering philosophy. More than 40 years ago, Taguchi faced many of the same kinds of situations Peter O'Brien would on the aerial refueling system guillotine mentioned at the beginning of this chapter. Taguchi recognized a need for an experimental approach that would extract statistically meaningful information from a minimized number of tests.

Based on the above, as well as other factors to be explained shortly, Taguchi developed a blend of engineering and statistical methodologies. These technologies emerged as a quality engineering philosophy and an approach to designing experiments that maximize information obtained from the experiment while simultaneously minimizing the test costs.

To more fully appreciate Taguchi's motivations, we must first understand Japan's background, and especially the position the country found itself in after World War II. These concepts have already been covered to a large extent elsewhere in this book. Here we need only recognize that the Japanese culture emphasizes maximizing process yield. Japan attempts to minimize scrap and waste at every opportunity. This was particularly true immediately after World War II, as the country's industrial base had largely been eliminated as a result of war-inflicted damage.

Taguchi formalized this approach through a management philosophy he described as the loss function, which is shown in Figure 11-9. In its simplest terms, the Taguchi loss function is based on variability reduction. Taguchi teaches that minimal variability in everything is inherently good, as discussed earlier in our reviews of statistics and statistical process control.

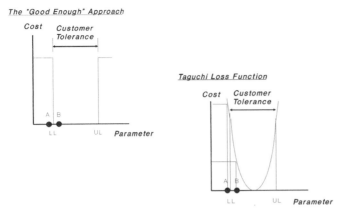

Figure 11-9. Taguchi Loss Function and Goalpost Quality Management Philosophy. Taguchi shows us that any departure from the best value represents a loss to society. His work emphasizes variability reduction. Under the outdated goalpost quality management philosophy, Point B is acceptable, but Point A is not. Taguchi teaches that there is not a fundamental difference between a point just inside specification limits (shown by Point B) and one just outside specification limits (shown by Point A).

Taguchi teaches that instead of simply trying to stay within specification limits, one should instead determine where within those specification limits the best value lies, and then devote the efforts necessary to minimize variability around that point. The Taguchi loss function emphasizes that departures from the optimal point represent a loss to society, and the larger the departure from the best value, the larger the loss to society.

Taguchi takes this concept of variability reduction even further. He totally rejects the so-called "goalpost" philosophy of quality management. In the goalpost philosophy, one recognizes that requirements have tolerances. Under the goalpost approach, any values lying within the allowed tolerance band are acceptable, and any lying outside the allowed tolerance band are unacceptable. Most of American industry has been managed according to the goalpost philosophy since the Industrial Revolution. This has led to the concept of meeting specification versus not meeting specification. The concept is the basis for much of what guides our country's quality control inspection function. Inspectors measure product characteristics to determine if they meet drawing requirements, or lie outside acceptable tolerance bands. The phrase "good enough for government work" undoubtedly emerged from this type of thinking.

Taguchi asks a fundamental question, and that is: What is the difference when using the goalpost approach to quality management between a dimension that lies just within the allowed tolerance band versus one that lies outside the allowed tolerance band? Taguchi maintains that the difference is insignificant, and based on that conclusion, some other approach to quality management must be pursued.

Think of the above in terms of Japan as an island nation, with its limited natural resources, and the underlying logic begins to make sense. If all of Japan's iron has to be imported to make steel, doesn't it make sense that anything that can be done to eliminate waste during the steel-making process minimizes the loss to society?

Taguchi had to determine an optimal blend of product and process characteristics to minimize production line loss and maximize product reliability (in line with the Taguchi loss function). Taguchi undoubtedly recognized that he did not have the luxury of performing large scale experiments with hundreds of test specimens to evaluate all potential process and product design parameters. Clearly, to minimize the loss to society during the engineering and manufacturing development process, a different approach was needed.

Taguchi developed just such an approach, which has been used extensively throughout Japan for at least two decades. Fortunately for us, the approach is also making great strides in the United States. The approach is generally called Taguchi design of experiments, and it offers a means for evaluating numerous design and process parameters with a minimum of test specimens. How powerful is it? Imagine being able to accurately evaluate the effects of seven variables with as few as eight test specimens. This chapter will review how this is done, but first, a few fundamental concepts must be understood.

Those High School Teachers Were Wrong!

Nearly everyone who will read this book has taken a high school science class, either in physics, chemistry, biology, or perhaps all three. With few exceptions, all of us were exposed to the so-called scientific method for conducting experiments. Without too much thought, we can recall our teachers explaining to us that a basic "scientific method" precept involved only changing one variable at a time when conducting an experiment. The thinking of that era (and unfortunately, the prevailing thinking for most American engineers today) was that if more than one variable changed, one would not know which of the changed variables was responsible for inducing any observed changes.

Let us examine the thinking that might have supported the above concept. For starters, most high school (and even college) laboratory course work does not involve physical rules influenced by more than one variable. If only one variable is at work, then, it makes sense that only one variable should be tested at a time. Also, most high school teachers (and far too many college professors) don't have the mathematics or statistics backgrounds to begin to evaluate the effects of more than one variable simultaneously.

Think about the problem from a different perspective. What really goes on in the world around us? Do we believe that most of the processes and products we use are only influenced by one variable, or that a change in one variable doesn't influence the effect of another variable?

Consider a fairly simple example. Suppose you want to determine which type of tire provides the best gas mileage for your car. You mount Brand X, for which the manufacturer recommends inflation pressures of 32 to 34 psi. You conduct the first set of trials with tires at 33 psi and record the gas mileage attained by your car. You next mount Brand Y tires, and you note that the Brand Y manufacturer recommends 33 to 35 psi inflation pressures. You conduct the next set of gas mileage trials. You keep the tires at 33 psi, because it's within the range recommended by the manufacturer and it's exactly the same as the pressure in the preceding set of trials. You don't want to change more than one variable at a time. After conducting the second set of mileage trials, you record the mileage, which is different than that for the first set of tires.

This presents a dilemma. Do we conclude, based on the results of your experiment, that one set of tires is preferable to the other? What if we had operated both sets of tires at different pressures? Would the other set of tires deliver better mileage under the influence of the different air pressures, and perhaps change your tire brand preference?

At this point, you may be thinking that the simple answer is to conduct additional trials at differing pressures for both brands, and then select the optimal blend of tire pressure and tire manufacturer. You could do this. In so doing, you recognize that one variable (tire pressure) interacts with the other variable (tire manufacturer) to produce a different outcome (gasoline mileage). Other questions emerge, however, that are quite troubling to the high school scientific method, and those who are accustomed to thinking on its terms. These questions include:

- How many different pressures should be tested?

- How many combinations of variables (tire manufacturer and tire pressure) will be required?

- Who will pay for all of this testing?

- What if the best combination of tire manufacturer and air pressures lies outside the air pressure recommendations of the tire manufacturer?

- Are there other things you need to be concerned about with the selected combination of tire pressure and air pressure (for example, handling, tire life, ride quality, and perhaps other characteristics affected by tire manufacturer and air pressure)?

The above questions put us in a real quandary. The example above is about as simple as it gets, yet the questions that emerge when we think about the effects of two variables influencing each other (as well as the outcomes of our simple experiment) are troubling. Clearly, the scientific method we learned in high school does not begin to answer any of the above questions (at least not in an economically viable manner). What would we do if we had to analyze a process or a design involving five or more variables, all of which could potentially influence each other, as well as several possible outputs? Stated differently, what would we do if we had a real world problem, just like the one facing Peter O'Brien at the beginning of this chapter? The simple facts are these:

- Most real world situations involve processes and product designs influenced by several variables.

- Combinations of variables frequently influence more than one output.

- The traditional scientific method, which emphasizes changing one variable at a time, is at best inefficient (due to the large sample sizes required to evaluate all variable combinations), and is at worst wrong (because it cannot evaluate the effects of variable interactions).

The bottom line? Those high school teachers were wrong! In today's world, we need a more efficient test technique. Ideally, we'd like to be able to evaluate the effects of several variables while only testing a small number of samples. We'd also like to be able to do this while varying several of the variables at the same time, as this will tell us if any interactions between the variables exist.

Where the Scientific Method Falls Apart

The above desire (i.e., modifying and testing several variables at the same time, and doing so with a minimum number of samples) runs completely counter to the traditional scientific method. The traditional scientific method implies that in order to determine the effects of several variables, we should only induce one change at a time. The traditional scientific method also implies that we should test all possible combinations of the variables (presumably, the concept is that testing all possible combinations will show if any particular combination is good or bad). This approach implies a very large sample size, as the number of possible combinations will rise exponentially as the number of variables

increase. Why exponentially? The number of possible combinations can be calculated using the formula:

$$N = L^V$$

where:

    N = number of possible test combinations
    L = number of test levels for each variable
    V = number of variables

If we wished to conduct a test to analyze the effects of 6 variables (each at two different levels), then we would have to test 64 possible combinations (2^6 = 64). That's a lot of testing, and usually, time and budget constraints won't allow such large sample sizes.

There is yet another problem with the above approach, and that is because even with our large sample size of 64 test specimens, we would only have a single data point for each possible combination. If a single combination of experimental variables passed or failed, would we really know anything? Recall our earlier discussion in Chapter 9 about the differences between deterministic and statistical thinking. We are faced with the same dilemma here, and the results could be just as misleading. Making a determination about the goodness of a product or process based on a single data point is not a good approach. Remember that all parameters will have a measure of statistical randomness, and the fact that a single test combination passes or fails does not necessarily indicate the particular combination of experimental variables will reliably work. What's needed is a different experimental approach that can evaluate the effects of several variables, with a small number of tests, to determine which have an influence and which do not.

Returning to Design of Experiments

We discussed shortfalls in the scientific method, and in particular, how relying on the success or failure of one data point could lead to misleading conclusions. The concern is that simply passing or failing a test does not necessarily reveal anything about the influence of any parameters we changed during the test, especially if the nature of statistical randomness is considered.

Suppose instead of simply passing or failing a test, we measure several test outputs and collect quantitative data (much the same as the weight data we dealt with earlier when performing our example analysis of variance). If we could perform ANOVA on the experimental results, would that not consider the effects of statistical randomness and help us to determine if any changes in the results were due to randomness alone, or perhaps to some variable we modified? The answer to this question is yes, and it forms the basis of the Taguchi approach.

Let us return to Peter O'Brien and the guillotine challenge he faces. In that situation, O'Brien is confronted with several potential variables, all of which affect the guillotine's ability to cut and clamp the hose. O'Brien has to determine which of the design parameters has a significant influence on guillotine performance, and design the guillotine to address these factors such that it will work reliably. His situation is an ideal Taguchi application.

The Taguchi Design of Experiments Process

Designing a Taguchi experiment is a seven-step process, as shown in Figure 11-10. These steps are explained in detail below, using the guillotine example described at the beginning of this chapter to illustrate the concept.

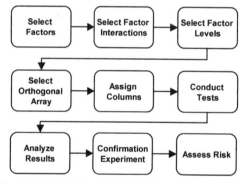

Figure 11-10. The Taguchi Design of Experiments Approach. The process selects factors and their interactions, develops an appropriate experimental design, analyzes results using the ANOVA technique, and concludes with a confirmation experiment.

Selecting Factors and Interactions

The first step in designing a Taguchi experiment involves selecting factors and interactions, or the

parameters to be evaluated during the experiment. There are several approaches for doing this, including brainstorming, flow charting, Ishikawa diagrams, and the systems failure analysis process described in Chapter 6.

The fault tree analysis approach presented in Chapter 6 is the most thorough for selecting factor interactions, and it was the approach O'Brien selected to converge on what might be causing the guillotine failures. O'Brien and his failure analysis team discovered that there were over 150 potential failure causes, but after preparing an FMA&A (as outlined in Chapter 6), the team ruled out all but three:

- *Cartridge Ignition Simultaneity.* The team knew that two cartridges were used to fire the guillotine, but they weren't sure both were firing simultaneously. They recognized the need to determine what effect nonsimultaneous firing might have on guillotine function.

- *Cartridge Lot.* Lot-to-lot cartridge output variability was also a potential cause of the guillotine failures. The team knew that such variability existed, but they didn't know if there was enough to influence guillotine function.

- *Ignition Current Level.* The guillotine cartridges are electrically fired, and one of the fault tree potential causes of failure included low current inputs into the cartridges. The team recognized a need to evaluate the effects of varying ignition current levels on guillotine performance.

The interactions between any of the above factors were not known, and O'Brien's team opted not to pursue this Taguchi capability while pursuing the experiment. That brought them to the next step.

Selecting Factor Levels

Selecting factor levels involves determining the values at which to set the factors during the experiment. The recommended approach is to select as many factors as possible during the initial Taguchi experiment, but to set these at a low number of levels (preferably only two). This will help to identify which factors make a difference, and subsequent experiments (if they are necessary) will help to establish exact design levels.

With the above in mind, how does one select factor levels? We recommend consultations with design engineers and using best case and worst case estimates of actual operating conditions to establish these levels. Here's what O'Brien's team selected:

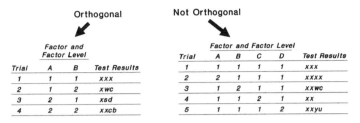

Orthogonal

Not Orthogonal

Trial	A	B	Test Results
1	1	1	XXX
2	1	2	XWC
3	2	1	XSD
4	2	2	XXCB

Factor and Factor Level

Trial	A	B	C	D	Test Results
1	1	1	1	1	XXX
2	2	1	1	1	XXXX
3	1	2	1	1	XXWC
4	1	1	2	1	XX
5	1	1	1	2	XXYU

Factor and Factor Level

Figure 11-11. The Orthogonal Array Concept. In an orthogonal array, each column has an equal number of samples at each level. The orthogonal array on the left is a typical Taguchi test matrix. The matrix on the right is not orthogonal. It represents a classical "scientific method" approach to experimentation, which changes only one variable at a time. The latter approach does not lend itself to ANOVA analysis techniques.

- For the cartridge ignition simultaneity factor, the decision was already made for the team. The ignition was either simultaneous or it wasn't. The team specified these two conditions for the experiment. Simultaneous ignition was denoted as Level 1, and nonsimultaneous ignition was denoted as Level 2.

- For the cartridge lot factor, the team selected an old lot and a new lot. New lot cartridges were denoted as Level 1, and older cartridges were denoted as Level 2.

- For the current level factor, the team found that the lowest level at which the cartridge would fire was 3 amperes, and the normal operating level was 10 amperes. The team selected 3 amperes (denoted as Level 2) and 10 amperes (denoted as Level 1) for the experiment.

These established the factor levels, which brought the team to the next step.

Selecting the Appropriate Orthogonal Array

Selecting the appropriate orthogonal array is one of the simplest steps in designing a Taguchi experiment. Taguchi made this easy for us. Taguchi texts have families of orthogonal arrays already developed, and selecting the appropriate one merely involves looking this up in a table in any Taguchi reference manual.

We've introduced a new term, orthogonal array, and we need to take a minute to define it. An orthogonal array is simply a matrix that has the same number of

conditions in each column. The concept is shown in Figure 11-11. This will allow for analysis of variance on the columns to determine which of the factors are significant, as we'll see shortly.

In selecting the appropriate orthogonal array, one need only know the number of factors to be evaluated, the levels of each factor, and the number of interactions to evaluate.

In O'Brien's situation, he wants to evaluate three factors (as described above), at two levels each. The appropriate Taguchi orthogonal array is the L8 (this is the designation by which Taguchi labels the orthogonal arrays). The L8 label merely means that the experimental design provides for eight tests, with each of the factors at two levels each. The orthogonal array shown in Figure 11-11 is an L4, and the one to be shown in Figure 11-12 (for this experiment) is an L8.

Assigning Factors and Interactions

Having selected the appropriate orthogonal array, the next step is to assign the factors to be evaluated to each column.

Taguchi provides detailed guidance on this. If the experiment is evaluating up to four factors, they should be assigned to the first, second, fourth, and seventh columns. The other columns are used for evaluating factor interactions, which is beyond the scope of what we will cover here (the reader seeking additional information is invited to review the texts listed in the reference section of this chapter).

	Column							
	A	B	C	D	E	F	G	Data
Trial	Ignition Simultaneity	Cartridge Lot	Not Used	Ignition Current	Not Used	Not Used	Not Used	Peak Pressure
1	1	1	1	1	1	1	1	7,140
2	1	1	1	2	2	2	2	7,560
3	1	2	2	1	1	2	2	7,980
4	1	2	2	2	2	1	1	7,560
5	2	1	2	1	2	1	2	7,000
6	2	1	2	2	1	2	1	7,280
7	2	2	1	1	2	2	1	7,700
8	2	2	1	2	1	1	2	7,490

Figure 11-12. The Taguchi L8 Orthogonal Array. This orthogonal array's columns have been assigned for evaluating three hypothesized guillotine failure causes. Each column represents a different factor being evaluated. The unused columns can be used for evaluating factor interactions, which is not addressed here. The 1 and 2 values in the array interior represent different levels for each factor.

Defining Test Specimen Configurations

The pre-defined Taguchi orthogonal array is one of the features that make the technique easy to use. Notice the "1" and "2" values assigned to the interior of the orthogonal array. All of the Taguchi orthogonal arrays have these factor levels assigned. After the factors have been assigned to the columns, one need merely examine the "1" and "2" values in the Taguchi orthogonal array and compare these to the factors and levels previously assigned to determine the configurations of each of the eight test specimens required for this experiment.

If we read across the orthogonal array in Figure 11-12 for the first test specimen, we see that cartridge ignition simultaneity is defined to be Level 1 (simultaneous ignition), cartridge lot is defined to be Level 1 (new cartridges), and ignition current level is defined to be Level 1 (10 amperes). The same can be done for each of the following seven test specimens to similarly define their configurations.

The Challenge: Determining What to Measure

Once the columns have been assigned and the test specimen configurations are defined, one is ready to begin the experiment. There's an extremely critical parameter yet to be defined, though, and that's what will be measured as an output variable (or, as it is often referred to, the response variable). So far, we've predicted which factors might make a difference in guillotine performance, selected different levels at which to test these factors, and filled in the appropriate orthogonal array. Analysis

of variance will be used to measure the effects of the different factors, but something has to be measured. Quantitative output data is needed.

This is one of the subtle but significant aspects of a Taguchi experiment. If we choose the wrong response parameter, the results of the experiment will be meaningless. We might have identified factors that make a difference, but if we do not measure the appropriate output, we'll never know.

How, then, does one go about selecting the appropriate output parameter to measure? We recommend defining a response that can be tied to the success of the device being tested. This takes some engineering intuition, and consequently, it's an area where engineering assistance is often needed.

There are also ways to mitigate the risk in selecting the output parameter to be measured. If there is some doubt about which parameter is more closely tied to the success of the device being tested, we recommend instrumenting all of the responses, and performing analysis of variance on each. When this occurs, one frequently finds similar conclusions from the analyses of variance performed on each output parameter, and that confirms that the results are meaningful.

For the guillotine Taguchi experiment, O'Brien selected peak pressure in the guillotine chamber as the output parameter upon which analysis of variance would be performed. This is also shown in Figure 11-12. One should also look for any attributes differences in performance (such as certain configurations failing and others passing) to help determine which parameters make a difference. Sometimes attributes data makes this very clear. Most of the time the analysis of variance is necessary to make this determination.

Performing ANOVA for a Taguchi Experiment

Once the output data has been collected, the next step is to perform the analysis of variance. This is fairly straightforward. We can use a simplified formula for the sums of squares for each of the columns:

$$SS_x = (\Sigma X1 - \Sigma X2)^2/N$$

where:

SS_x = the sum of the squares for each column

$\Sigma X1$ = the sum of the values in the column at Level 1

$\Sigma X2$ = the sum of the values in the column at Level 2

N = the total number of specimens (in the guillotine case, this is eight)

For O'Brien's Taguchi experiment, the sum of the squares for ignition simultaneity is calculated below:

$\Sigma X1 = 7,140 + 7,560 + 7,980 + 7,560 = 30,240$
$\Sigma X2 = 7,000 + 7,280 + 7,700 + 7,490 = 29,470$
$SS_A = (\Sigma X1 - \Sigma X2)^2/N = (30,240 - 29,470)^2/8 = 74,112.5$

We may similarly calculate results for the other factors to find that the sum of the squares for the old versus the new cartridge lots is 382,812.5, and the sum of the squares for ignition current levels is 612.5.

Analyzing the Results

Armed with the results of the Taguchi experiment, O'Brien now knows that with the current guillotine design, cartridge lot makes the biggest difference. Ignition simultaneity also plays a role, although it is not as significant as which cartridge lot is used. Ignition current level, based on the experimental results, appears to have inconsequential effects. What does all of the above mean?

To O'Brien, it means that he has to redesign the guillotine to be insensitive to cartridge lot and ignition simultaneity. These are the factors that make a difference with the present design, so the design has to be modified so that they do not have an impact on guillotine performance. This concept of modifying the design so that it is insensitive to parameters like these makes the design robust.

The Confirmation Experiment

After completing all of the above, Taguchi recommends repeating the experiment to confirm the results. We concur with this recommendation, especially if the design is being modified to make it more robust. There are risks associated with Taguchi testing, but the confirmation experiment helps to mitigate these.

Summary

The traditional experimental approach of changing one variable at a time and evaluating the results is time consuming, expensive, and often misleading because it fails to consider the effects of normal statistical variability.

ANOVA is an analysis technique that helps to overcome this limitation by looking for differences in average performance, and then making a comparison to determine if these differences are due to normal statistical variation, or some combination of statistical variation and other special influences.

Taguchi takes the ANOVA concept several steps further, and offers a family of designed experiment templates for evaluating the effects of several factors with small numbers of test specimens.

References

Introduction to Quality Engineering, Genichi Taguchi, Asian Productivity Organization, 1986.

Taguchi Techniques for Quality Engineering, Philip Ross, McGraw-Hill, 1988.

Quantitative Techniques for Management, Thomas Payne, Reston Publishing, Incorporated, 1982.

Chapter 12

Quality Function Deployment

Understanding and satisfying customer expectations...

Art Levenson, the PAVE VIPER laser chief engineer at Omega Lasers, was confused. The engineers in his conference room were arguing strongly about which features of the PAVE VIPER laser should predominate in design tradeoffs, and which requirements should take a back seat to others. The PAVE VIPER laser engineering development program was just getting started, and the development program was scheduled for completion in twelve months. At the very beginning of the program, though, there were already serious disagreements within Levenson's team on which features of the laser should be emphasized.

"The Air Force has already told us the energy levels they want," Tom Axelson, Omega's optical physicist, said. "We have to meet that requirement."

"But if we meet that, we'll suffer elsewhere, and we think we have to hold the weight down," Bill Olson, the mechanical designer, answered. "We are suffering with a set of conflicting requirements. Another poorly thought-out wish list from a customer who doesn't really know what he wants..."

Art Levenson listened to the conversation unfolding in front of him as he thought about the PAVE VIPER program. The PAVE VIPER laser transmitter/receiver would be part of the larger PAVE VIPER system used on U.S. Air Force tactical aircraft. The PAVE VIPER system was being developed as an externally mounted, underwing system containing complex sensors and navigation systems. PAVE VIPER would provide its host aircraft with terrain-following capabilities to allow the aircraft to fly nap-of-the-earth in

conditions of total darkness. PAVE VIPER would also provide target acquisition (using television and thermal imaging sensors) and target designation capabilities.

For target designation, the PAVE VIPER system would use a laser transmitter/receiver. The target designation capability would be met by the laser illuminating the target with a laser beam, and then providing a source of reflected laser energy from the target for a new generation of smart munitions to home in on.

As the chief engineer on the PAVE VIPER laser effort at Omega, Art Levenson was charged with designing a laser receiver with capabilities well beyond any developed to date, including higher laser energy output, lighter weight, higher reliability, and a provision for eye-safe training with friendly ground troops. Levenson knew that all target-designating military lasers operated on the 1.06 micron standard tactical wavelength. The 1.06 micron wavelength was hazardous (it blinded humans if viewed without special filters). The hazardous nature of the laser beam prevented meaningful training in combined arms exercises, and because of this limitation, the Air Force wanted to add an eye-safe feature to the laser that allowed it to transmit with a beam using a 1.54 micron wavelength that was not hazardous to the human eye.

Levenson knew he could make an eye-safe laser by adding another set of optics in the PAVE VIPER laser to convert the beam from 1.06 microns to 1.54 microns, and designing the laser to switch the beam from one optical path to another on demand. Doing so, however, added weight and complexity to the laser transmitter/receiver. Levenson knew that he faced a severe weight constraint, and added complexity usually meant reduced reliability simply because there would be more parts that could fail. What characteristics would be most important to the customer?

Levenson knew that the Air Force wanted the PAVE VIPER laser to operate at higher energy levels than previous military lasers. The Air Force wanted greater lasing range, and to allow the laser to work more effectively with smart munitions in fog or smoke. Providing higher laser energy output required pumping more energy into the laser, and it also required a larger-capacity laser cooling system. Both of these features would add weight to the laser. The higher energy levels would also be tougher on the laser's internal optics and would reduce the laser's reliability.

"Look," said Tom Axelson, with such feeling that it shook Levenson from his thoughts, "these guys don't know what they want. We're going to have to tell them. And if they don't like it, they'll have to learn to live with it, because we simply can't give them everything they want without designing a laser that's bigger than the airplane it's going on."

"That's not the right approach," Levenson said. The rest of the engineers and physicists in the room fell silent. Levenson continued. "If there are conflicts in what the laser has to do, we can't make the tradeoffs independently and then tell the customer to take it or leave it. Our company got where it is today by being responsive to the customer. We're the number one developer and producer of military lasers in the world. We're in that position because our customers prefer us, and they prefer us for a lot of reasons. One reason is our technical expertise, which comes from the people in this room and others across the hall and out in the factory. Another reason, though, is that we meet the customer's needs. Our problem here is that this is the first time the customer has levied a set of requirements on us that seem to be in conflict with each other. Our challenge is to put our minds into this job, with the customer, and determine which requirements are most important and which requirements can be relaxed."

Quality Function Deployment

Quality Function Deployment (or QFD, for short) is a basic TQM tool that systematically develops customers' needs and expectations. The tool provides a graphical methodology for unearthing a customer's stated and unstated needs and expectations, for making decisions in cases where these needs and expectations conflict, and for driving these customer-based requirements and expectations into the product development and manufacturing process. QFD is driven by what the customer wants, and for this reason, the technique is often described as "deploying the voice of the customer."

QFD originated in the late 1960s and early 1970s in Japan, when the Mitsubishi Corporation developed it for defining shipbuilding requirements at the Kobe Shipyards. Building ships involves enormous expenses and a small number of products. For these reasons, Mitsubishi recognized the importance of determining in great detail exactly what their ship-buying customers wanted before beginning the design process.

Mitsubishi recognized many other factors that could influence their ship-buying customers' needs and expectations. Potential conflicts between customer-expressed requirements (the situation described earlier on the PAVE VIPER laser program) would influence customer needs and expectations. Requirements the customer might not express (or perhaps might not even recognize, but requirements the customer would want satisfied nonetheless) would influence the design process. These requirements might include expectations so basic they might not have even entered the customer's requirements-listing thought process. It almost goes without saying that ships shouldn't leak. Because this is such an obvious requirement, a watertight hull would probably not be listed as a customer requirement, but the requirement exists. Another group of requirements the customer might list include such things as government regulations, and perhaps other externally imposed requirements.

Mitsubishi achieved notable success satisfying the needs and expectations of their customers using the QFD approach. Based on Mitsubishi's success, the approach soon caught on in other industries in

Japan. Toyota used the technique in developing automobiles, and from Toyota it spread to the American automobile industry (most notably Ford Motor Company). QFD is now making headway in other industries, including defense, aerospace, and other high technology areas. QFD began as an engineering tool to assure that the development process resulted in a product meeting consumer needs and expectations, but because it does so, it also provides strong marketing advantages to those organizations who choose to use it.

Integrating Other Activities

QFD is a planning tool that guides the application of resources to meet customer needs and expectations. As mentioned immediately above, QFD provides marketing advantages in that it allows companies to develop and offer products that are responsive to the needs and expectations of their customers. This is particularly true when new products are offered, as new products developed using QFD are far less likely to miss the requirements and expectations of their target market. This allows companies using QFD to be first with products that exactly meet their customers' needs and expectations. Capturing a market with a new product responsive to customer needs is easier than doing so with one that is not.

Capturing market share is also easier if the product is the first of its kind (as opposed to attempting to capture market share once someone has already developed the market). QFD supports getting a new product to market ahead of the competition. Why is this? Because QFD assures that products are developed in response to customer needs and expectations, and doing this requires the integration of several key organizations within a company (most notably, Engineering, Manufacturing, and Marketing). This interdisciplinary integration helps to eliminate surprises when the product moves from design and development to production, and then from production to the customer. In the process of working as a team to develop a QFD analysis, Engineering designs and develops a product that is responsive to market needs and can be efficiently manufactured. This occurs because Marketing and Engineering must work closely together to assure that the product meets customer needs and expectations. Engineering must also work closely with Manufacturing in preparing a QFD, and that

provides early feedback on product features that detract from producibility. A more producible product emerges. Fewer post-product-introduction changes are likely, and products can typically be developed and offered to the customer in less time than would be possible if the QFD approach was not used.

QFD can also be used to improve existing products by redefining customer needs and expectations. This may be necessary as a result of evolving customer requirements, or perhaps because the product did not do well initially in meeting the needs of the customer. Although the technique can be used after products are developed, it is best applied early on in the development process. Changes implemented after a design is committed to production are far more expensive than changes made prior to production startup (post-production-startup product redesigns require tooling changes, process changes, documentation changes, and cost-inducing changes).

Another important QFD advantage is the preservation of knowledge related to the underlying reasons for product design features and customer needs and expectations. Once a QFD matrix has been prepared, it can be easily reviewed as the product evolves, as new product developments are initiated, and as new people join the organization. QFD matrices prevent reinventing the wheel where customer needs and expectations are concerned.

QFD Prerequisites

Quality Function Deployment is a fairly advanced TQM concept, and as such, cannot be implemented in an organization that does not have other underlying TQM concepts in place. These prerequisite underlying TQM concepts include recognizing that most quality problems are systems related (i.e., they are not caused by workmanship errors), a company culture that emphasizes working as a multi-disciplinary team, an organization's willingness to invest more time up front in the product development process to prevent fewer problems downstream, and using statistical process control and design of experiment technologies.

W. Edwards Deming teaches that most quality problems are not the result of workmanship problems, but are instead due to poorly defined processes and management systems (Deming postulates that only 15% of all problems are due to people problems; the other 85% are due to poor management systems, product deficiencies, or process deficiencies). Accepting this concept is necessary in order to successfully implement the QFD approach. If an organization does not believe it must create a product design, a process, and a management system that eliminate much of the potential for quality deficiencies, the QFD team members will be unable to successfully deploy the voice of the customer to create a product that can consistently meet customer needs and expectations.

As explained above, multi-disciplinary involvement and a willingness to have engineers, marketers, and manufacturing personnel work together to deploy the voice of the customer is also a prerequisite for successful QFD analysis. The product design must be responsive to the needs and expectations of the customer, and it must be producible. Good interdepartmental communications are necessary, particularly between an organization's management and the engineering and marketing departments. Engineers and management must recognize that customers buy an organization's products (the engineers probably buy very few), so what the engineering department thinks is important has to take a back seat to what the customer wants. This is hard for many engineers to accept, but successful companies are market-driven (not engineering-driven).

QFD is by nature a preventive tool, and like most preventive tools, it requires that an organization be willing to invest earlier in a program than it otherwise would (i.e., without QFD). Many organizations focus on problem detection instead of problem prevention. Organizations instill these cultures without realizing it.

Consider the people in most organizations who are rewarded. Successful people in many organizations are typically good "fire-fighters" (i.e., people who resolve crises). The problem with "fire-fighters" is that they almost never focus on fire prevention (and in effect, they become arsonists, creating the sparks that will start the next fire). QFD helps to reverse the opportunities for this form of organizational arson.

Defining the needs and expectations of the customer early in the product development process will help to prevent problems by creating a product that meets customers' needs and expectations. Working through the QFD process with both the engineering and manufacturing departments will help to assure producibility. Both of these QFD features help to prevent products that customers don't want, or that have high reject rates during production. The bottom line here is that in order for QFD to work, organizations have to be willing to make the investment early. The time spent on QFD early in a program can result in significantly less time spent later detecting and correcting problems (problems that might not have occurred if QFD had been performed).

Statistical process control is necessary to assure product consistency. Both statistical process control and design of experiments technologies (in particular, Taguchi techniques) are necessary to assure that variability is minimized. Variability reduction is inherent to acceptable quality, and meeting customer needs and expectations simply cannot be accomplished if products are allowed to vary widely in areas important to the customer. Stated differently, if important product attributes cannot be optimized and controlled, the conclusions of a QFD matrix will be of little use to anyone.

QFD Mechanics

Quality Function Deployment is a graphical analysis technique that portrays customer needs and expectations, how these needs and expectations will be satisfied, and tradeoffs between conflicting needs and expectations. The analysis uses what at first appears to be a complex matrix often referred to as the house of quality. Although the matrix appears complex, once the technique is understood the matrix actually simplifies the presentation of a large amount of information.

There are six basic information elements in the QFD matrix:

- What the requirements are.
- How they are to be met.
- Relationships between the requirements and how they are to be met.

- Target values for the requirements.
- Relationships between how the requirements are to be met.
- A quantification of the importance of the requirements.

As an optional feature, QFD matrices can also be expanded to benchmark a product against competitor products.

Each of the QFD matrix elements is further developed below.

Defining Requirements: QFD WHATs

The WHATs are the QFD definition of customer needs and expectations. These are listed in a column that appears on the left side of the QFD matrix, as Figure 12-1 shows.

WHAT

| high reliability |
| light weight |
| high energy |
| 1.06 wavelength |
| 1.54 wavelength |
| one technician |
| standard tools |
| no fastnr tools |

Figure 12-1. Preliminary PAVE VIPER QFD Matrix. The PAVE VIPER QFD begins by showing the WHATs, or the Air Force requirements and expectations.

Let's return to the PAVE VIPER laser discussed at the beginning of this chapter. Art Levenson, the PAVE VIPER chief engineer, knows from his conversations with the Air Force and his prior experience in developing lasers that the customer's requirements include:

- High reliability.
- Light weight.
- A high laser energy level.
- A 1.06 micron tactical wavelength.
- A 1.54 micron "eye safe" training wavelength.

As mentioned earlier, not all customer requirements and expectations are expressed. Does the above list include all of the customer requirements and expectations? In many instances, customer expectations must be developed through market surveys (most often used for commercial consumer goods) or through extensive interviews and requirements development meetings with the customer (this approach is most often used for institutional procurements of complex items, like the PAVE VIPER laser).

Levenson had several meetings with the Air Force, and based on his probing of their needs and expectations, he learned that there are other expectations the customer did not express in their initial set of specifications for the PAVE VIPER laser. Levenson knew that earlier Air Force laser systems were extremely difficult to maintain, and the Air Force had moved to systems with lasers that could be removed or installed by a single technician. Levenson also knew that the Air Force flight line technicians disliked the special fasteners used to secure the lasers in the earlier systems. The flight line technicians preferred current generation laser systems with bolt designs that did not require special tools, but could instead use ordinary socket wrenches. Based on these unstated customer requirements, two more requirements should be added to the QFD matrix list of WHATs:

- Removal or installation by one technician, and

- Installation hardware requiring only standard tools.

The unspoken requirements are ones that can be particularly damaging if not addressed. Unspoken WHATs are essentially requirements that are so obvious to the customer that they are taken for granted. If met, these unspoken needs do not particularly increase customer satisfaction. If not met, though, the unspoken needs can do a great deal to decrease customer satisfaction.

There is yet another category of WHATs, and those are the quality surprises. Quality surprises are features customers do not expect, but like and need once these "requirements" are met. Stated differently, these requirements are neither spoken nor unspoken; they are unrecognized features which, once the customer is exposed to them, become recognized requirements. Vanity mirrors on automobile sun visors are good examples (few people thought to ask for them, but automobile manufacturers started including the feature, and now everyone expects it).

Levenson had an idea to use toggle screw fasteners to secure the PAVE VIPER laser (these would require no tools at all, but could instead be manipulated manually). Levenson therefore decided to add another requirement to the QFD matrix list of WHATs:

• Fasteners requiring no tools at all.

Figure 12-2 shows all of the above WHATs in a preliminary QFD matrix.

Meeting Requirements: QFD HOWs

Once the list of WHATs is determined, the next step in performing QFD analysis is to determine how each WHAT can be achieved. This leads to the QFD HOWs, which convert the WHATs into actions designed to satisfy the WHATs. The HOWs are also known as substitute quality characteristics, as they are substituted for the WHATs. The HOWs (or substitute quality characteristics) are what an organization has to implement in order to satisfy the WHATs. The first step in defining the HOWs involves simply listing each of the WHATs described above, and describing HOW each will be met. Most people prefer to first do this in a side by side listing, as Figure 12-2 shows.

Let's return to the PAVE VIPER laser discussed above to consider a few of its HOWs. One of the WHATs (high reliability) is translated into HOWs that require low laser energy levels and a low optics count. Both of these features will help to improve laser reliability. Another one of the WHATs (high laser energy level), when converted into a HOW, requires high levels of electrical energy pumped into the laser. Note that this HOW and its associated WHAT (high laser energy levels) is in direct conflict with low laser energy levels (which is one of the HOWs required for high reliability). Carrying this further, we see that another WHAT is the 1.54 micron training wavelength, and that its associated HOW is a parallel optics path. This parallel optics path mandates the use of additional lenses, prisms, and mirrors. This HOW (i.e., additional optical

elements) creates a situation that is in direct conflict with another one of the HOWs related to high reliability (the HOW that requires a low optics count).

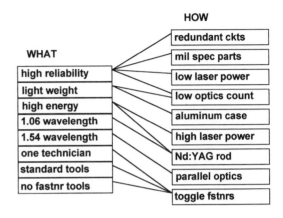

Figure 12-2. Converting WHATs to HOWs. This list shows how the QFD WHATs will be satisfied. The HOWs are often referred to as substitute quality characteristics.

What the above situation shows is that sometimes the HOWs can affect more than one WHAT, and these effects are often in conflict. A HOW that supports one WHAT can detract from another WHAT. To cope with this situation, one needs a capability to identify and evaluate interactions between the HOWs. Such a capability would allow for tradeoffs between cases where conflicts arise.

A first step in making these resolutions is to create the QFD matrix by listing the HOWs across the top of the matrix, and the WHATs on the left side of the matrix. Figure 12-3 shows that portraying the WHATs and HOWs in this manner creates a matrix with rows (representing the WHATs) and columns (representing the HOWs). The matrix format allows one to show the manner in which all of the HOWs affect each of the WHATs. This is graphically portrayed through the use of symbols that are placed at the intersections of the rows and columns in the QFD matrix.

During early QFD development by Mitsubishi in the Kobe shipyards, Mitsubishi used horse-racing symbols for win, place, and show from a local racetrack. These are shown in Figure 12-3. Concentric circles denote a strong correlation (in other words, the HOW strongly supports attaining the WHAT for their intersecting column and row). A simple circle shows a medium correlation (the HOW moderately supports attaining the WHAT where the column and row intersect). A triangle shows a weak correlation (the HOW only weakly supports attaining the WHAT where the column and row intersect). In many cases, there will be no relationship between a WHAT and a HOW, and in these cases, their respective column and row intersections are left blank.

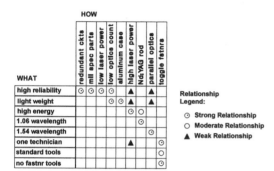

Figure 12-3. Expanded PAVE VIPER QFD Matrix. This expanded matrix adds HOWs to the QFD, which shows how the PAVE VIPER WHATs will be satisfied. The symbols inside the matrix show the relations between the WHATs and the HOWs, as explained in the relationship legend.

After listing the WHATs and the HOWs on the left side and on the top of the QFD matrix, one proceeds to evaluate each of the column and row intersections to assign a concentric circle, a circle, a triangle, or no symbol to evaluate the relationship between every WHAT and HOW. This is shown in Figure 12-3 for the PAVE VIPER laser QFD.

Having completed the QFD matrix to this point, one should pause to analyze the emerging QFD findings. These findings can fall into any of several categories. A column with no entries, for example, means that a HOW has been incorporated into the product that does not satisfy any of the WHATs. Stated differently, the product may have a feature that serves no purpose. In such cases, it might be wise to consider deleting the feature to reduce cost (one should do this with caution, however, as the HOW may be addressing a WHAT the QFD preparers neglected to consider). Another case might involve a situation in which a QFD row is blank. This indicates a customer need or expectation (one of the WHATs) is not being

addressed by any of the HOWs. The message to the product development team is that the design concept must be altered to satisfy the WHAT. The objective of performing the QFD review at this point is to make sure the design concept addresses all customer needs and expectations, but does not include features with no value. The goal is product responsive to what the customer wants.

Quantifying Expectations: HOW MUCH

Having identified the QFD WHATs and HOWs, the next step requires quantifying these expectations by defining HOW MUCH. The design team needs to know what constitutes satisfaction of a customer expectation or need, and this is accomplished by identifying HOW MUCH for each HOW. Defining each of these requirements in the form of a QFD HOW MUCH allows the product development team to define success, and then ascertain if the emerging design is a success.

Returning to the PAVE VIPER laser example, the HOW MUCH requirements can be defined by listing each of the HOWs, and then determining HOW MUCH of each HOW is required to meet customer requirements. This determination is accomplished through customer surveys, engineering analysis, customer specifications, and comparison to other competing products (or benchmarking, as will be discussed below). In the QFD matrix, the HOW MUCH entries are listed along the bottom of the matrix, as shown in Figure 12-4.

Putting on a Roof: The Correlation Matrix

When the engineers at Toyota began to use QFD as a tool to speed the product development cycle and create cars more responsive to customer needs, they recognized frequent conflicts between QFD HOWs.

When such conflicts are discovered, one has to decide which takes precedence and to what extent. Resolving a requirements conflict by prioritizing the requirements is referred to as a tradeoff (one requirement is traded off against another to arrive at the compromise most acceptable to the customer). Toyota's engineering staff developed an optional addition to the QFD matrix to assist in performing these tradeoffs. This addition, which appears at the top of the QFD matrix as shown in Figure 12-5, is called a correlation matrix.

HOW

WHAT	redundant ckts	mil spec parts	low laser power	low optics count	aluminum case	high laser power	Nd:YAG rod	parallel optics	toggle fstnrs
high reliability	⊙	⊙	⊙	⊙		▲		▲	
light weight			⊙	⊙	▲		▲		
high energy				⊙	○				
1.06 wavelength						⊙			
1.54 wavelength							⊙		
one technician					▲			⊙	
standard tools								○	
no fastnr tools								○	
	2 ckts	all	< 50	< 70 lb	> 100 J	one	2 paths	all	

HOW MUCH

Figure 12-4. PAVE VIPER QFD Matrix Showing HOW MUCHs. The QFD has been further developed to quantify the HOWs.

Figure 12-5 shows that the QFD correlation matrix is made up of diagonal lines that allow comparing one HOW to another. The concept is to show if one HOW supports or is in conflict with another HOW. The form of the correlation matrix is such that it allows making this comparison between all of the HOWs. Recall that in the previous discussion regarding the symbols inside the QFD matrix, the comparisons were being made amongst the effects of a HOW on a WHAT. In the correlation matrix, the comparison is being made strictly among the HOWs.

Staying with the PAVE VIPER laser example we can see some conflicts among the HOWs. Low laser power and high laser power have a strong negative correlation, as does a low optics count and parallel optics paths (which would add optics to the laser). The correlation matrix also shows a moderate conflict between the redundant circuits and low laser power (redundant circuits would require additional power to the laser).

As described earlier for the central portion of the QFD matrix, symbols are also used for the correlation matrix. These symbols were developed by Toyota and are used to show the degree of

correlation amongst the HOWs (the symbols are shown in Figure 12-5). Note that the correlation matrix symbols are different than the symbols used for the central portion of the QFD matrix. In the QFD correlation matrix, two concentric circles are used to show a strong positive correlation, one is used to show a positive correlation, an "x" is used to show a negative correlation, and a double-barred "x" is used to show a strong negative correlation.

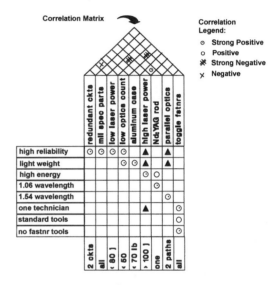

Correlation Matrix

Correlation Legend:
⊙ Strong Positive
○ Positive
✖ Strong Negative
× Negative

	redundant ckts	mil spec parts	low laser power	low optics count	aluminum case	high laser power	Nd:YAG rod	parallel optics	toggle fstnrs
high reliability	⊙	⊙	⊙	⊙		▲		▲	
light weight			⊙	○	▲		▲		
high energy				⊙	○				
1.06 wavelength						⊙			
1.54 wavelength							⊙		
one technician					▲			⊙	
standard tools								○	
no fastnr tools								○	
	2 ckts	all	< 50	< 70 lb	> 100 J	one	2 paths	all	

Figure 12-5. PAVE VIPER QFD With Correlation Matrix. The correlation matrix adds a roof to the house of quality. Its purpose is to assist in analyzing tradeoffs between the HOWs.

Once the correlation matrix is complete, it allows for a rapid comparison of the HOWs. In particular, the correlation matrix highlights conflicts between the HOWs. Conflicts typically develop due to physical constraints, violations of the laws of physics, customer desires, and perhaps other factors.

Conflicts are resolved through the tradeoff process. When tradeoffs are required, they are most typically accomplished by making adjustments in the HOW MUCH area. These decisions (which HOW MUCH to increase, which HOW MUCH to decrease, etc.) are based on customer requirements and expectations. Since the tradeoffs may require compromises of certain customer requirements or expectations, engineering requirements, management intervention, and consultations with the customer frequently guide such decisions.

Another QFD Option: Benchmarking

The concept of benchmarking was introduced a few pages ago. Benchmarking is a process that compares a product to competing products in order to assess its comparative strengths and weaknesses. This analysis is also known as a competitive assessment.

There are two kinds of product benchmarking: one driven by inputs from the company's customers (or potential customers) and another performed by the company itself (usually by its engineering department). Both are recommended when performing a QFD. Why perform both benchmarking analyses?

Certainly, benchmarking by a company's customers is of paramount importance, as such an analysis provides tangible feedback on how well customers perceive products meet their needs and expectations. Benchmarking by the company's engineering department is also useful, as it provides useful guidance to a company on how its product stacks up against the competition.

More significant, however, are the differences that often emerge between an internal benchmarking exercise and one based on customer inputs. An organization's engineers may feel that their product fares well against the competition, when customers feel exactly the opposite. Engineers may feel that their product has a significant advantage, but customers may see the engineers' perceived advantages as not being particularly significant. Customers may recognize strong advantages in competing products' features that a company's engineers have not yet recognized. The two benchmarking analyses (i.e., the internal and external benchmarking exercises) serve as an important reality check for an organization. The combination of these two analyses assure that an organization is market-driven rather than engineering-driven. Remember that quality consists of meeting customers' needs and expectations. It's the marketplace that buys products (not the engineers who develop them).

Customer competitive benchmarking is primarily accomplished through the use of customer surveys. The results of the survey data are included to the right of the QFD matrix, as shown in Figure 12-6. To simplify the presentation, symbology is used, with a different symbol representing each company whose products are being compared. Typically, a square is used for the company developing the product, and circles, triangles, and other symbols are used for competitors. When the customer competitive assessment is included in the QFD, one side of the assessment is labeled "good" and the other side is labeled "bad," with the different companies' symbols positioned between these extremes. A set of symbols for each company is positioned in the customer competitive assessment along each of the WHAT rows. This shows how well customers feel the product satisfies each of their needs. Figure 12-6 shows this for a customer competitive assessment based on Air Force inputs concerning PAVE VIPER laser performance.

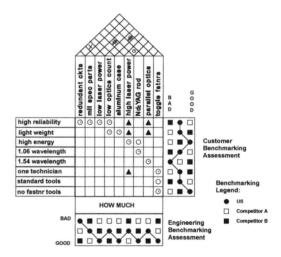

Figure 12-6. PAVE VIPER QFD With Benchmarking Analysis. The QFD now includes a comparison of Omega's lasers to those of its competitors, both from customer and Omega engineering perspectives. A careful review of these two benchmarking analyses will show that Omega's engineers need to reconsider their opinions.

Internal competitive benchmarking, as mentioned above, is similar to customer competitive benchmarking except that it is performed by the company developing the product. In most cases an organization's engineering department will perform this analysis, perhaps with assistance from the company's marketing group.

The internal benchmarking results are presented in a

manner similar to the customer competitive benchmarking results, using the same symbology, except that they are shown below the QFD matrix directly opposite the HOWs (see Figure 6). Recall that customer needs and expectations are presented as WHATs in the QFD matrix, and the customer competitive benchmarking results are shown in a manner that allows categorizing responses with respect to each of the WHATs. The internal competitive benchmarking results are similarly shown at the bottom of the matrix, arranged against their respective HOW categories. Remember that the HOWs represent how a company satisfies each of the WHATs, so it's appropriate that the internal benchmarking results be portrayed against corresponding HOWs, while customer benchmarking results be portrayed against corresponding WHATs.

Once both sets of benchmarking results are included in the QFD matrix, one should again search for the contradictions between the internal and customer benchmarking analyses discussed above. If a customer thinks a need is not being satisfied, but the engineering department believes that it is (or that it is not important), something needs to change. An example of this situation is illustrated in Figure 6. If one examines Figure 6, it becomes obvious that Omega's engineers have a higher opinion of their product than does the customer.

Quantifying QFD Matrices

In addition to the techniques described thus far, the importance rating concept is another approach that helps to further increase the usefulness of the QFD analysis. The importance rating concept consists of weighing each of the WHATs, assigning numerical values to the symbols used in the center portion of the QFD matrix, and then multiplying the two to give a quantified values referred to as importance ratings.

Weighing the WHATs is usually performed by the customer (companies performing QFD simply ask the customer to assign numbers to each WHAT to indicate its importance to them). The WHAT weighing values are often shown between the WHAT listing and the central portion of the QFD matrix (see Figure 12-7). The symbols in the interior portion of the QFD matrix are weighted by

assigning nine points for a strong relationship, three points for a medium relationship, and one point for a weak relationship.

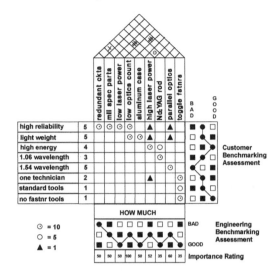

Figure 12-7. PAVE VIPER QFD With Importance Ratings. On the left side of the matrix are weighing values determined by the customer for each of the WHATs. The relationship symbols are each rated at 10, 5, or 1 point as shown above. The value of the symbols in each column is multiplied by the customer-assigned WHAT weights. The resulting values for each column are summed to provide the importance ratings shown at the bottom of the matrix. These values assist in making tradeoffs between conflicting requirements.

The results of these importance rating multiplications are shown at the bottom of the QFD matrix, below the HOW MUCH values, as depicted in Figure 12-7. The importance rating analysis for the PAVE VIPER laser, using hypothetical WHAT weightings developed with the customer, are also shown in Figure 12-7.

When using the importance rating concept, it's important to recognize that the importance values have no meaning in and of themselves. These values only provide a tool for evaluating the relative importance of each WHAT and HOW to the overall quality of the product, or how each should be considered when evaluating the product's overall quality. The importance rating concept also provides another vehicle for performing tradeoffs when WHATs or HOWs conflict (i.e., when the QFD matrix shows a negative correlation). Figure 7 shows such a situation. The low optics count has a strong negative correlation with parallel optics, as mentioned earlier. How do we decide which is more important? The low optics count importance rating is 100, compared to the parallel optics importance rating of 60. This tells us that based on the customer's needs and expectations, the low optics count should receive a higher priority (we should be willing to sacrifice some or all of the parallel optics path).

Summary

This chapter presented the QFD analysis approach and several QFD advantages. These include identifying customer needs and expectations, determining how to meet them, defining quantified goals, and methodologies for identifying and resolving conflicting requirements. One of the advantages of QFD analysis is that it deploys the "voice of the customer," and forces product development teams to focus on customer needs and expectations (not just what the developers feel is important). As mentioned earlier, QFD is a relatively advanced TQM concept, and is probably best employed when used in conjunction with other previously-implemented TQM disciplines (quality measurement, Taguchi design of experiments, statistical process control, and others).

The reader should also recognize that although this chapter focused on new product development in an engineering environment, the QFD technique is equally at home on product improvement efforts, and even when developing a manufacturing approach. Organizations that use QFD successfully prepare numerous matrices for the product's concept development phase, detailed design work, and various phases of the product's manufacture. The approach is the same: Focus on the customer's needs and expectations, and develop everything else in a manner that optimally satisfies these needs and expectations.

References

"Implementing Internal Quality Improvement With the House of Quality," K.N Gopalakrishnan, B.E. McIntyre, and J.C. Sprague, *Quality Progress*, September 1992.

"Total Quality Management, Part II: TQM Team Training," *TRW Systems Integration Group Training Materials*, January 1990.

Quality Function Deployment Awareness Seminar, Ford Motor Company Quality Education and Training Center, 1989.

Chapter 13

Inventory Management

Reducing inventory to improve quality...

Ray Jiminez looked at the annunciator circuit card and shook his head. The avionics digital bus analyzer Jiminez was attempting to assemble failed three acceptance tests in a row. Each time, the avionics analyzer provided an Error Code 163 message, which meant that the circuit card Jiminez now held in his palm was defective.

When the first failure occurred, Jiminez relied on the avionics bus analyzer's built-in test equipment and he replaced the annunciator board. The avionics analyzer promptly failed again. On the second failure, Jiminez thought he might be dealing with a failure elsewhere in the avionics analyzer, and he pulled the second defective board to test it in a known good avionics analyzer. The good avionics analyzer failed, too, confirming that the built-in test equipment properly isolated the fault to the circuit card.

Jiminez looked at the annunciator board. It was the third defective circuit card the materials handlers had delivered to him from the stock room. He could see nothing wrong with it, but he knew that electronics failures were often not visible to the naked eye, and he had to rely on what the avionics analyzer's built-in test equipment told him.

Jiminez shook his head. Perhaps the entire inventory of annunciator cards was bad, but then he remembered that Digital Technology was near the end of the production run on the avionics bus analyzer, and there would only be a few cards left in the stockroom. Jiminez knew that there had been Error Code 163 problems with the annunciator card in the past, but there had always been enough in inventory to simply pull another board (and that had always allowed the system to pass).

Jiminez thought of the bin for the annunciator cards in the stockroom. He could visualize it, nearly empty, with perhaps three or four circuit boards remaining. He wondered if the cards that were left would work.

Inventory Is a Vital Part of Industry

With the onset of interchangeable parts and mass production techniques, carrying huge inventories of raw materials and partially-completed products (often referred to as work in process, or WIP) became a normal part of industrial management in America. Until the last two decades, these huge inventories were regarded as necessary and perhaps even desirable. American manufacturing management recognized the delays believed to be inherent in setting up to manufacture new products. Our managers used huge production runs to minimize the effects of setup-induced delays and their associated costs. If the setup costs were spread over the production run, larger runs would naturally result in a smaller per unit setup cost. Larger inventories would also build in a safety margin (or so the prevailing philosophies held) such that if there was a sudden increase in consumer demand, the demand could be more expeditiously met. Larger inventories were also thought to offer a buffer against quality problems. The conventional wisdom for most of our history put forth the notion that defectives found in work in process could simply be replaced by other products pulled from inventory, without interrupting the production line.

This inventory management philosophy carries a heavy burden, and that burden is the cost of carrying the inventory. What do we mean by the cost of carrying the inventory? Somebody has to pay for the inventory, pay for warehousing the inventory, and pay interest on the money used to do both. Money tied up in paying for the above is money that cannot be used for other purposes. Who pays for all of this? The company carrying the inventory, and that means the cost is ultimately passed on to the customer.

Based on the above, it seems to make sense that organizations would be interested in minimizing their inventories in order to minimize costs. Companies have been interested in doing that ever since the industrial revolution, but the approach initially pursued by American industry (and therefore the world, at least for most of this century) focused on the wrong problems. Let's see why.

Managing Excess Inventory

For most of America's manufacturing history, industry has followed the economic ordering quantity (or EOQ, for short) philosophy of inventory management. This approach is centered on the notions discussed above (i.e., that large work in process and other inventories are desirable for satisfying unpredictable consumer demand and providing a safety net for quality deficiencies). The approach emerged in the early 1900s, and it has endured. Today it is even taught in American business schools.

EOQ is a mathematical technique that considers customer ordering patterns, the costs associated with placing orders and carrying inventories, and supplier delivery times. The underlying concept is that EOQ determines ideal quantities of raw materials and other elements of work in process to satisfy customer demands without carrying too much inventory. What constitutes too much inventory? From the perspective of a company using the EOQ approach, it's any amount beyond that which is reasonably required to meet anticipated customer demand.

Here's how the approach works. It's a two-step process that begins by determining the optimal ordering quantity and optimal average inventory level by using EOQ analysis. This technique allows a company to determine how much inventory it should order and maintain in an idealized situation (a situation in which delivery times and demand

levels remain constant). Under idealized conditions, EOQ analysis defines the quantity and inventory level necessary to minimize inventory carrying costs while providing sufficient inventory to meet customer demands.

The next step involves determining the safety stock. The safety stock adds a little margin above what would be required for the ideal situation in which delivery times and demand remain constant. The safety stock recognizes that delivery time and demand are typically not constant, and in fact, the safety stock is derived based on the history of prior delivery times and demand levels.

The EOQ approach to managing inventory became fairly sophisticated over the last 60 or 70 years. Safety stocks are determined by statistically analyzing the distributions of prior delivery times and demand levels. The cost of money (i.e., the interest rate) also affects the determination. The cost of warehousing the excess material is factored in. Finally, the fixed costs related to ordering materials are included.

Is there anything wrong with the above approach? Absolutely. The EOQ approach provides a very sophisticated answer to a fundamentally flawed philosophy of inventory management. It's based on several assumptions, most of which have to do with the premise that delivery times cannot be reduced and reserve quantities are necessary to make up for changes in customer demand or other factors (such as in quality problems and rejections).

What's needed is a paradigm shift to help us see that delivery times can change, customer demand can be met with a responsive manufacturing system, and inventory reserves aren't necessary to make up for quality shortfalls.

Just-In-Time Inventory Emerges

Industrialized nations depend heavily on steel, aluminum, copper, plastics, and other materials. These raw materials use petroleum products heavily. Some have a petroleum content (like the plastics) and all require petroleum in the machinery used to manufacture products. The industrialized nations of the world are highly dependent on oil, and they have been for some time.

In the early 1970s, the world's oil producing nations started to restrict their oil exports to drive up prices. Many of us are old enough to remember the panic that ensued in the United States. Gasoline prices shot up from 30 or 40 cents a gallon (where it had been for at least the previous ten years) to over a dollar a gallon (where it has remained ever since). Long lines at the gas pump, previously unheard of, became the norm for several months. States enacted temporary legislation to restrict gasoline consumption through the use of approaches that limited when one could purchase fuel. One common scheme involved license plates (those with plates ending in odd numbers could buy gasoline on certain days, while those with license plates ending in even numbers could buy gasoline on other days). We were in a desperate situation, and we tried to legislate and politicize our way out of it.

In the United States, the above approaches did not last long. The 1970s oil shortages were basically a contrived situation. In reality, the shortages were nothing more than the exporters restricting supplies to drive up demand and prices consumers were willing to pay for gasoline. Our nation's politicians realized that the resulting long lines at the gas pumps and consequent restrictions on driving were intolerable to the vast majority of Americans, and unless a solution emerged, these politicians would soon find themselves voted out of office. To be sure, America enacted many steps to conserve fuel. We developed more efficient automobiles and initiated other moves to cut back on the consumption of petroleum products. The American response to the Organization of Petroleum Exporting Countries' oil embargo was essentially a political solution to provide Americans an unrestricted supply of gasoline and other petroleum fuels (at a much higher price, which was what the oil exporting nations wanted when they restricted the oil supply).

While Americans found political solutions to the oil embargo, the Japanese took the opposite approach. As discussed in previous chapters, Japan has few natural resources. The Japanese culture emphasizes a minimalist approach, and Japan searched for ways in which to reduce their dependence on imported petroleum. In the early 1970s (when the oil embargoes began), Japan already had small, fuel efficient vehicles. Where else could they save fuel?

Japanese leaders knew their industry consumed large amounts of fuel, and they searched for ways to reduce this consumption. The Japanese did not want to reduce their output; they only wanted to increase their efficiency by maintaining or increasing output while decreasing the amount of fuel they consumed.

The Japanese knew the production of finished goods consumed large amounts of petroleum, and that the oil embargoes of the early 1970s were potentially devastating. Instead of focusing on a political solution to the artificially-induced oil shortages, however, the Japanese instead focused on approaches to reduce their oil consumption. The Japanese determined that one means of realizing this goal was to reduce their work in process inventories without reducing final output. The Japanese concluded that they had to use manufacturing processes with little or no scrap (which they were already doing, as described in other chapters), and they had to reduce their work in process inventories to the bare minimum required for meeting their production schedules.

Let's consider the last part of the above statement. Reducing the work in process inventory seems to be about what American management had been attempting to do all along with the EOQ and safety stock approaches discussed earlier. After all, the intent was to assure the minimum amount of inventory was kept on hand to reliably meet consumer demand while allowing for variations in demand, supplier delivery times, and production lead times (i.e., the amount of time required to set up and manufacture goods).

The fundamental flaw in the above approach, though, was that supplier delivery times and production lead times were assumed to be rigid. In other words, the economic ordering quantity and safety stock approach to inventory management assumed these two parameters were inviolate, and with those two flawed assumptions, Americans determined their minimum acceptable work in process inventories.

The Japanese took a different approach. They questioned fundamental assumptions about supplier delivery times and production lead times. The Japanese recognized that if supplier delivery times

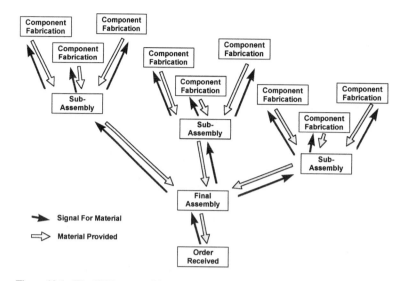

Figure 13-1. The JIT Inventory Management Approach. The system is based on a "pull" concept, where each work center signals the preceding work center as it requires work in process inventory. There are no work in process queues. Raw materials (not shown here) would also be ordered only as needed.

to produce the finished products from the preceding work station. These "pulls" for work in process inventory cascade from the final assembly area back through the factory. The concept is illustrated in Figure 13-1.

The basic concept is elegantly simple. As a final product is needed, the final assembly area notifies the work stations providing it with the sub-assemblies required to produce the final assembly. Prior to this signal from the final assembly area, and in an ideal JIT environment, the subassembly areas would have no inventory. The subassembly areas make only the required amount of subassemblies to fill the order for the final assemblies, as signaled by the final assembly area. When the subassembly areas are notified that they need to provide subassemblies, they signal their supplying component fabrication areas (or the purchasing organization, if the necessary components are procured items). The idea is that in an ideal JIT environment, there is no inventory other than a very small work in process inventory, and the work in process inventory only consists of that needed to satisfy current orders. There are no stocks of completed parts, subassemblies, or finished products in a JIT environment.

The Japanese call this process of signaling for materials as needed a "kanban" system. "Kanban" means "signal" in Japanese. One of the advantages of a signal-oriented system is that it is inherently simple. Complex computerized inventory management systems are not necessary. Work centers only need to send signals to their supplying work centers elsewhere in the factory when materials or subassemblies are needed. In many factories that have moved to a JIT inventory management approach, the signals are quite simple. One can send specially-designed containers for only

and production lead times could be reduced to zero (i.e., if suppliers could deliver instantly when asked to do so, and internal production lead times could similarly be reduced to nothing), then the requirement for a work in process inventory could be greatly decreased. Greatly decreasing the work in process inventory would tie up less cash, and require less oil for use in raw materials and the operation of internal factory equipment.

With that realization, the Japanese then set about developing approaches to reduce factory inventories to zero, and to only receive materials and work in process when needed. The essence of this inventory management approach is to receive materials at each stage in the manufacturing process "just in time" to initiate the next manufacturing operation.

The Basics: Just-In-Time Inventory Management

The just-in-time (or JIT, as it is more commonly known) inventory management philosophy is centered on the concept of providing materials only when needed at each work station throughout the factory. The system is often referred to as a "pull" system, as it starts with an order for finished products and each work station pulls what it needs the amount of materials that are needed, raise color-coded flags, or simply verbally inform the supplying work centers which items (and how many) are needed.

Making JIT Work

Most American managers are familiar with inventory management approaches that provide for manufacturing items throughout a production facility in a coordinated manner such that all of the components and subassemblies are available when needed to support the overall end item delivery schedule. Lead times, supplier delivery times, and other factors are considered by inventory and production control managers in a non-JIT environment to back into required production schedules.

Managers accustomed to thinking this way are naturally wary of the JIT concept when first exposed to it. Why this natural aversion? Because managers accustomed to thinking in conventional terms cannot divorce their thinking from the lead time concept, and most inherently believe some fallout (or scrap) in the work in process inventory will necessitate building a safety stock.

What's needed, then, to make the JIT concept work? Two things: rapid setup times (to eliminate the lead time phenomena that drives up work in process inventory), and little or no scrap (to eliminate the need for safety stocks of work in process inventory). Let's take a look at each of these fundamental JIT building blocks.

The Consequences of Setup Time

Most of the lead time inherent to the production process in a non-JIT environment is created by what is known as queue time and setup time. Queue time is how long work in process waits at each work center until it is worked on. Setup time is the time required to modify or adjust manufacturing machinery for the work to be performed. In most instances, the actual run time (or the period of time production operations are actually performed) is quite small compared to queue time and setup time.

The signaling for material inherent to the JIT approach naturally eliminates the queue time. The

material simply is not delivered to the work center until it is needed, and it's not needed until it will actually be used by the work center.

If the queue time is eliminated by signaling for materials as needed, only the setup time is left as the principal contributor to lead time in a JIT environment. Accordingly, reducing the setup times becomes extremely important if a JIT inventory management approach is to work.

How big a contributor is setup time? Most people don't recognize the significance of this productivity detractor. Even manufacturing engineers and managers don't recognize the losses incurred as a normal part of production operations due to machine setup, especially if they have not been exposed to JIT inventory management concepts. In a conventional production operation (i.e., a non-JIT environment), machine setup times of hours and sometimes days are not uncommon. The natural fallout of long setup times are a tendency to want to make as many parts or subassemblies as possible once the setup is complete (thereby aggravating the excess work in process inventory problem), and the unavoidable scrap that results as nonconforming parts are produced in attempting to set up a production area to manufacture a particular item.

Implementing JIT requires reducing setup times. Most successful JIT operations have areas in which setup times have been reduced from hours to seconds. Obviously, when setup times are reduced in this manner throughout a facility or production process, the process becomes much more responsive to downstream demand, and converting from manufacturing one part or subassembly to another becomes much less of a chore. An organization that reduces its setup times moves closer to a JIT capability.

What is involved in reducing setup times, and how does one go about doing so? The process usually involves identifying those features of a machine that have to be adjusted, and then determining how to do so rapidly. If a milling machine uses a screw adjustment to move the cutting wheel to different locations for different products, it might make sense to use spacers for each product. When production shifts from one product to another, all the operator needs to do is insert the milling machine spacer for the new product (instead of tediously using the screw adjuster, making a single new part, scrapping the setup piece if the part is nonconforming, readjusting the machine to try another part, etc.). Such setup reduction activities can make for dramatic throughput and capacity increases.

Normally, the setup reduction approach would be assigned to the manufacturing engineering organization. This makes sense, as manufacturing engineering is typically responsible for process development and optimization. In many cases, though, we've observed that the operators assigned to a production area are the best source of ideas if the setup time reduction challenge is presented to them. After all, the operators know what they have to go through to set up manufacturing equipment.

Lowering the Water Level

Along with setup time reduction, processes that yield little or no scrap are required to make JIT work. Why is this? Consider the JIT concept, and its fundamental underlying principle of providing materials only as needed to reduce work in process. What happens if any of the intermediate process steps produce only the amount of material needed, and then at a subsequent operation, defective parts are discovered? The nonconforming items have to be scrapped or reworked and the operation slows down. In an attempt to produce material to make up for the nonconforming items, more material is added to the work in process inventory. This works against the JIT concept.

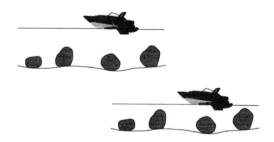

Figure 13-2. Lowering the Water Level. Moving to JIT is like lowering the water level in a lake with submerged boulders. As the water level drops, the boulders become visible and the need to eliminate them becomes apparent. The same situation exists for nonconforming hardware as inventory levels are reduced. Formerly hidden nonconformances become obvious, as does the need to eliminate them.

An interesting phenomenon occurs when companies start to implement the JIT management concept. Invariably, companies begin by reducing their work in process inventories. In most cases, the work in process inventories contain nonconforming materials. When the work in process inventories were larger than needed to support delivery requirements, the nonconforming materials were tolerated. This is not to say that the nonconforming materials were used in producing hardware delivered to the customer. By "tolerated," we mean this: Organizations not using JIT knew if defective materials were discovered, there would probably be enough good material on hand to replace the nonconforming material. What happens as inventories are reduced and there isn't enough good material on hand? At this point, a company has two choices:

- The company can return to its former inefficient ways and simply keep larger stocks of work in process on hand (and hopefully have enough to replace bad material with good as the bad material is discovered).

- The company can eliminate the nonconformance root causes such that it only has good material in its smaller work in process inventory.

Obviously, the second option is the preferred one, as it allows the company to continue to reduce its work in process inventory. This process of work in process inventory reduction, nonconformance discovery, and nonconformance root cause elimination occurs in just about every JIT implementation.

Many observers compare it to lowering the water level in a lake. There might be many rough boulders in a lake that threaten passing boats, but unless one lowers the water level, they remain invisible. As Figure 13-2 shows, when the water level is lowered, the boulders are visible, and they can be addressed.

Summary

American inventory management philosophies have been based on the economic ordering quantity concept, which factors in production lead times and variabilities in customer demand. This approach results in large work in process inventories. The JIT inventory management approach works to reduce lead times to near-zero values such that when demand exists, setting up to meet it takes almost no time. This allows a responsive production system, but only if the system produces quality hardware. To make the JIT concept work requires eliminating nonconformance root causes. As inventory levels are reduced, previously hidden nonconformances become visible and their causes can be attacked. Companies that implement JIT inventory management can expect increased profitability from lower inventory carrying costs and fewer nonconformances.

References

Zero Inventories, Robert W. Hall, Dow Jones-Irwin, 1983.

Financial Analysis on the IBM-PC, Joseph and Susan Berk, Chilton Book Company, 1984.

Japanese Manufacturing Techniques, Richard J. Schonberger, The Free Press, 1982.

World Class Manufacturing Casebook, Richard Schonberger, The Free Press, 1987.

Well Made In America, Peter C. Reid, McGraw-Hill Publishing Company, 1990.

Chapter 14

Value Improvement

Deleting cost and adding quality...

Tom Marino showed visible disappointment as he reviewed Alpina's monthly sales figures. As the marketing vice president for Alpina Bicycles, Marino was responsible for meeting aggressive sales goals. From the report in front of him, Marino could see that unless he could implement a significant change, Alpina would not have a successful year. There was still time to recover (the report showed January sales figures, traditionally a slow month for bicycles), but knew that he needed to implement a major change in marketing strategy. Marino asked his secretary to convene a meeting with his three marketing managers, Ed Smith, Al Buddingsworth, and Elaine Reiswig. The four met in Marino's office.

"What's the problem here?" Marino asked. "All three of you show significant declines in your areas."

The group sat silently for several seconds. Finally, Reiswig spoke. "Everyone knows we make a top quality mountain bike. We were the industry leader, but we lost the lead. Our product is overpriced compared to the competition," she said. Before Reiswig could continue, Buddingsworth interrupted.

"Overpriced significantly," Buddingsworth said. "We used to be very cost competitive in our target market, but times are tough. People are turning to less expensive bikes. It's that simple."

Smith spoke next. "Our target group recognizes our superior quality, but they don't want to pay the premium that goes with it. We're at least $80 higher than our competition on almost every model. Our challenge is going to be to keep our reputation for

building a high quality bicycle and get our pricing more in line with the competition. People care about quality, but these days they expect it, and they don't want to pay a premium for it."

Marino was surprised by the unanimity of opinion. He looked out the window, lost in thought. How in the world can I get our costs down without adversely impacting quality, he wondered...

Understanding and improving an organization's cost structure is far more complex than one might initially imagine. In many ways, the challenge is quite similar to that discussed in Chapter 4, which discussed quality measurement. In that chapter, we constructed a quality data system to answer a basic question:

Where should we focus our quality improvement efforts?

The challenge in cost reduction is similar, but it is also more complex. In cost reduction, one not only has to find out where the costs are (no simple task in itself, as any chief executive will readily admit); one also has to find out which costs can be eliminated or reduced without adversely affecting quality.

There is a difference between cost reduction and value improvement. Cost reduction is simply what its name implies: the reduction of costs. Value improvement goes considerably beyond cost reduction. Value improvement reduces cost while satisfying or exceeding customer expectations.

Consider the issue of value improvement from a broader perspective, and in particular, its relation to

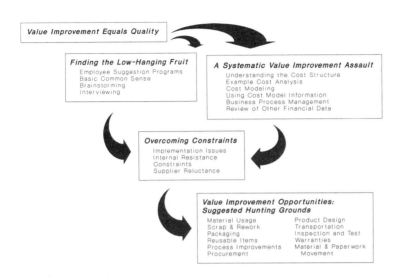

Value Improvement Equals Quality

Finding the Low-Hanging Fruit
Employee Suggestion Programs
Basic Common Sense
Brainstorming
Interviewing

A Systematic Value Improvement Assault
Understanding the Cost Structure
Example Cost Analysis
Cost Modeling
Using Cost Model Information
Business Process Management
Review of Other Financial Data

Overcoming Constraints
Implementation Issues
Internal Resistance
Constraints
Supplier Reluctance

**Value Improvement Opportunities:
Suggested Hunting Grounds**
Material Usage
Scrap & Rework
Packaging
Reusable Items
Process Improvements
Procurement
Product Design
Transportation
Inspection and Test
Warranties
Material & Paperwork
Movement

Figure 14-1. Basic Value Improvement Elements. The process first grabs the readily apparent value improvement opportunities (the low-hanging fruit). It continues with a systematic assault on unnecessary cost.

quality. Why is value improvement being discussed in a book on quality? One could argue that value improvement might be more appropriate in a book on financial management, or perhaps a book focused on the subject of cost reduction.

The linkage between quality and value improvement is simultaneously obvious and subtle. To understand the issue, turn to the definition of quality, or perhaps more appropriately, a definition of poor quality. What is poor quality? Poor quality is anything that does not meet expectations, and therefore results in dissatisfaction. The essence of value improvement is the ability to maintain or exceed customer expectations while removing cost. Doing so more thoroughly satisfies customer expectations (receiving the same level of service or product satisfaction at reduced cost can only serve to better meet customer expectations). If customer expectations are more thoroughly satisfied, has quality improved? The answer is yes.

This concept runs counter to most traditional thinking. If costs are reduced (or so the conventional wisdom runs) the product or service is being cheapened. Unfor-tunately, this is the typical result of many cost cutting efforts. All of us have

seen evidence of this, as illustrated by:

- The automobile manufacturer that turns to cheaper materials (resulting in a less solid "feel" and reduced customer satisfaction).

- The stereo manufacturer that uses lower grade components (resulting in lower sound quality, inferior product reliability, and reduced customer satisfaction)

- The home construction company that installs cheaper carpeting (resulting in a less durable home and reduced customer satisfaction)

All of the above examples involve cost reduction. Cost reduction (if done imprudently or without regard to customer expectations) will cheapen a product or service. Value improvement reduces costs, but not at the expense of customer expectations. When cost is removed from a product or service without compromising the satisfaction of customer expectations, from the customer's perspective both quality and value improve.

The linkage between value improvement and quality improvement is also consistent with the teachings of Dr. Genichi Taguchi. Taguchi's approach to quality (as discussed in the previous chapter) is based on the premise that anything for which quality has not been optimized represents a loss to society. If the loss to society is minimized through value improvement, does quality improve? Again, the answer is yes, as we will show throughout this chapter.

Two Value Improvement Approaches

There are two basic methods for identifying and eliminating unnecessary cost (see Figure 14-1). One involves a systematic assault aimed at identifying and eliminating unnecessary cost drivers (many of which are quite subtle). The other is a simpler approach that involves attacking the more obvious cost drivers. Let's discuss this approach first.

Grabbing the Low-Hanging Fruit

Many value improvements are obvious (see Figure 14-2), and require no special analyses or other techniques to unearth. These are the cost elements that are clearly unnecessary, have no redeeming value, and are often glaringly apparent (at least they appear to be after having been discovered). Going after this low-hanging fruit is the way most organizations approach identifying and eliminating unnecessary costs. And often, after having done so, this is where they stop, too. This is unfortunate. As we'll see later in this chapter, there's often a lot left on the tree. For now, though, let's concentrate on finding the more obvious cost drivers. Methodologies for identifying these cost drivers include employee suggestion programs, brainstorming sessions, interviews, and basic common sense.

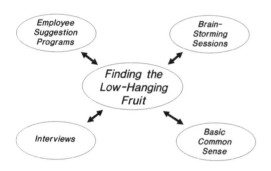

Figure 14-2. Grabbing the Low-Hanging Value Improvement Fruit. This approach involves employee suggestion programs, brainstorming, interviews, and basic common sense.

Employee Suggestion Programs

Much of the discussion in this book and effective management in general revolve around the concepts of listening well, and employee involvement and empowerment. Suggestion programs, covered earlier in Chapter 7, offer a structured approach for doing both. A value improvement suggestion program is simply a formalized mechanism that allows organizations to ask employees for ideas on how to eliminate unnecessary cost. Our experience shows that four factors are essential to assure the

success of a suggestion program: making it easy to submit suggestions, publicizing the program, incentivizing the process, and getting answers back to all participants (whether the suggestion is incorporated or not).

Making it easy to submit suggestions is deceptively difficult. We've seen programs fail simply because the suggestion forms were too complicated, or because the submittal process was made unnecessarily intimidating. This can occur if the program requires the suggestor's manager to review and approve the change, or if elaborate cost justifications from the suggestor are required. Bear in mind that we live in a society where a significant percentage of the population is functionally illiterate. A larger percentage has reading and writing skills only slightly beyond illiteracy. In many cases, these people constitute a sizable portion of the factory work force. We believe most of the people with this handicap are painfully aware of and embarrassed by it. Most of the unfortunate folks with this deficiency will go to extreme lengths to avoid anything that might reveal their illiteracy. That doesn't mean they don't have good ideas on how to eliminate unnecessary cost, however. The bottom line here is that any suggestion program requiring pre-reviews or complicated explanations and analyses will cut off a lot of good ideas. Our advice is to require no reviews prior to the submittal of suggestions, and to do away with complicated suggestion forms. Simply ask for the suggestions. Some people will provided their ideas on paper. Others will come forward to present their ideas verbally. We've even seen companies offer to provide assistance in submitting the suggestion.

Incentivizing suggestion programs by offering fiscal rewards greatly increases the number of suggestions, particularly if the rewards are financially significant, paid promptly after the suggestions are approved, and well publicized. Our observation is that an award equal to 1 percent to 3 percent of the savings resulting from an employee suggestion can be extremely significant. If an employee offers an idea that saves a company a million dollars, giving $10,000 to $30,000 is not unreasonable. Our suggestion of 1 percent to 3 percent is small enough that it won't be missed by the organization (particularly when compared to the savings), and will undoubtedly be appreciated by the recipient.

Our experience confirms that attempting to reduce costs by offering small awards is counterproductive. If someone finds an approach for saving a million dollars and is rewarded with $100 or even $500 for the effort, the reward will probably be regarded as an insult.

Publicity is equally important in insuring the success of a value improvement suggestion program. Company newsletters, posters, meetings, and word of mouth all help. Unusual publicity approaches can also support these programs well, especially if they are tied to the product. One company that manufactures fuel tanks uses a prominently displayed "Suggestion Tank" with excellent results. Publicizing awards is also a good idea. When employees see a peer receive a check for $15,000 or $20,000 from the company's chief executive, everyone's creative juices begin to flow. This is substantiated by the sharp rise in numbers of suggestions that typically follow a publicly awarded cash bonus.

Our experience confirms that answering every suggestion is a requirement. We believe it's equally important to respond to all suggestions immediately to inform those submitting the suggestions that their ideas have been received and are being considered. It may take weeks or months to evaluate the acceptability of a suggestion, so letting people know that their ideas have been received and are being considered eliminates any potential indication of nonresponsiveness. We have learned that it's also important to let those whose suggestions are not implemented know the reasons why. Sometimes the person who made the suggestion can refute the logic that led to disapproval. If the people who disapproved the suggestion are wrong, they need to know (why pass up potential savings?).

Brainstorming

There are several excellent books and articles available on brainstorming, so we won't spend time here developing the brainstorming concept. The concept works well, and should be considered in any value improvement program. In fact, this chapter is based on a series of brainstorming meetings that lasted for approximately two months. The sessions were designed to take unnecessary cost out of a major production program, and they achieved significant results (the value improvements exceeded $10 million). The results and methodologies that emerged from the meetings were distilled, categorized, and documented to guide our thinking on future value improvement activities.

A format we have seen work well is to approach a value improvement brainstorming session in the following manner:

- Announce the meeting and its purpose at least three days in advance (this gives the participants time to start thinking about the subject).

- Prominently display three elements of information throughout each meeting: The purpose of the brainstorming session (for example, eliminate unnecessary cost from the Acme widget), the product's requirements (those things that cannot be compromised), and the inviolate rule of brainstorming (there is no such thing as a stupid idea).

- Invite no more than six participants, and assure different departments are represented (e.g., engineering, manufacturing, procurement, marketing, quality assurance, etc.). This will help to pull out more ideas, as a manufacturing person will know of different cost elements than does an engineer, a marketing person may be aware of costly features for which customers have no use, etc.

- Designate a recorder, and list the ideas on either a chalkboard or poster paper (this helps to stimulate others' thinking, and provides a record of the ideas for future development or discussion.

- Develop a list of action items from each meeting (specific tasks for members to accomplish prior to the next meeting), and work these between meetings in disciplined manner.

When conducted well and pursued vigorously, brainstorming invariably ends up following the logic trail developed in the next section of this chapter, which is a systematic approach to value improvement. Brainstorming can also provide a checklist of target-rich value improvement areas.

To facilitate our readers' efforts, these are included near the end of this chapter.

Interviews

The people who have the best insight into potentially unnecessary costs are those closest to the operation. This includes shop floor personnel (both manufacturing technicians and quality assurance inspectors) and first-level supervision. We've learned a great deal by simply asking these people what portions of their jobs make no sense, or what they would eliminate if they had the power to do so. The answers to the above questions are often embarrassingly obvious. In one case involving a munition system, an assembler showed us that the manufacturing instructions required:

- Applying a lubricant in one operation.

- Removing the lubricant after two unrelated operations

- Cleaning the lubricated area with solvent such that paint (applied in the next step) would adhere.

- Re-applying the lubricant after the painting operation.

The application and removal of the lubricant (and the cleaning operation prior to painting that the lubricated area required) were unnecessary. The initial application of lubricant most likely originated during an earlier version of the manufacturing process, but subsequent revisions to the process added steps that made lubricant application and removal unnecessary. This probably occurred as a result of the process engineer not fully understanding the process, or making changes to one portion of the process without being aware of actions required elsewhere.

In this situation, many managers and manufacturing engineers had observed the redundancies for years without recognizing what they were seeing. Yet the assembler (the person closest to the work) knew that the process had evolved into a wasteful one. She knew, because she was the person doing the unnecessary work. She only needed to be asked.

Basic Common Sense

Many people can observe situations and immediately recognize value improvement opportunities (the elimination of redundant inspections or tests, unnecessary paperwork, etc). Most of us regard this ability as little more than basic common sense. Unfortunately, common sense really isn't all that common. We've seen literally thousands of instances similar to the example mentioned above. Once obvious unnecessary cost drivers are identified and eliminated, people joke about the "ignorance" that created and allowed the conditions to exist. Many of the people who joke about these are same ones who observed the conditions for years without recognizing the waste. Here's the message: If your organization is blessed with people who possess the "common" sense to immediately recognize value improvement opportunities, by all means take advantage of their talents. But don't be disappointed if such wisdom is not immediately forthcoming. Sometimes the only avenue available is a systematic assault on unnecessary costs, and that's what the remainder of this chapter is all about.

A Systematic Value Improvement Assault

The discussion above presented techniques for finding obvious waste, and for pulling all employees into the value improvement process. This search for obviously unnecessary cost (the low-hanging fruit) can provide high returns, but taken alone, it won't maximize the return from an organization's value improvement efforts. The low-hanging fruit analogy is particularly appropriate here. If only the above approaches are used, one tends to leave a good deal of value improvement fruit hanging on the tree. With the aid of a few other tools (the techniques we'll present below), one can greatly increase the harvest. This takes our quest for value improvement to another level: a systematic assault on cost to uncover value improvement opportunities. Such an approach is outlined in Figure 14-3. A systematic value improvement assault involves establishing objectives, identifying costs, prioritizing costs from several perspectives, and evaluating the necessity of each. This requires understanding the cost structure, and then analyzing each element of the cost structure for its contribution to customer satisfaction.

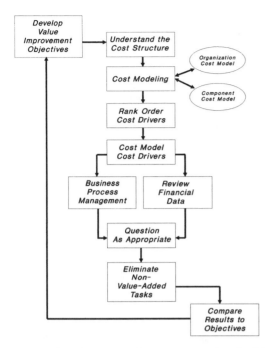

Figure 14-3. A Systematic Value Improvement Assault. This approach involves developing value improvement objectives, and then systematically identifying and evaluating the need for each cost contributor.

Understanding the Cost Structure

Understanding an organization's cost structure is not an easy task. Lee Iacocca once explained that even Chrysler didn't fully understand how much it cost to build a car. Improving value requires systematically stripping away unnecessary costs, and to do that, you have to know three things: what the costs are, where the costs are, and which costs are unnecessary.

Knowing where the costs are constitutes the first challenge. To meet this challenge, we recommend the use of cost models based on organizational and system component costs.

Let's return to Tom Marino at Alpina Bicycles. Marino's challenge is to improve the value of Alpina's bikes by reducing cost without sacrificing quality. Marino has made this the basis of his marketing strategy to improve the Alpina's sales. Let's examine Alpina more closely to see how this can be accomplished.

Alpina Bicycles buys most of its bikes' components from other suppliers, and assembles them in its plant. To support Marino's new marketing strategy, Alpina wants to improve the value of its most popular model, the Rockchucker. Marino wants to do this by improving quality and lowering cost in a move to posture the Rockchucker as the mountain bike value leader, and thereby improve the company's market share. Figure 14-4 shows Alpina's costs (as it understands them) for the Rockchucker mountain bike.

Component	Cost	Supplier	Department
Wheels	$ 45.37	Yakima	Procurement
Brakes	23.62	Sunstrate	Procurement
Gears	52.68	Omnigear	Procurement
Saddle	18.54	Pacer	Procurement
Handlebar	9.07	Track	Procurement
Pedals	11.11	Track	Procurement
Crown	6.89	Track	Procurement
Frame Tubing	5.40	Yakima	Procurement
Tires	7.45	Genroad	Procurement
Paint	3.37	Pacco	Procurement
Assembly	8.13	Alpina	Manufacturing
Inspection	2.27	Alpina	QA
Total:	193.90		

Figure 14-4. Alpina Rockchucker Mountain Bike Costs. This table shows the cost of each component in the Alpina Rockchucker mountain bike, and whether the component is manufactured by Alpina or made by an Alpina supplier.

The first challenge Marino faces in evaluating Alpina's cost structure is identifying the Rockchucker's costs, and where these lie. Determining which costs are unnecessary can be attacked from several perspectives, but to organize the cost reduction process, Alpina must first evaluate where the costs are concentrated.

This evaluation is best performed through an

identification of the costs (as shown in Figure 14-4), and through the use of displays that organize the costs by component and organization. We recommend a pie chart format, as outlined in Figure 14-5.

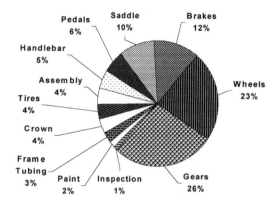

Figure 14-5. Component Cost Model for the Alpina Rockchucker Mountain Bike. This model shows that 71 percent of the bike's costs are concentrated in four components.

As can be seen in Figure 14-5, 71 percent of the costs are concentrated in the 4 most expensive components (the gears, wheels, brakes, and saddle). Accordingly, it would make sense to evaluate the cost makeup of these components first, based on the assumption that those items with the highest cost probably offer the greatest potential for cost reduction. This may not always be true, as will be discussed later, but as a starting point, the assumption makes sense.

Organizational cost models can be prepared in any of several manners. One might be interested in assessing all costs, in which case it makes sense to assess both supplier and internal costs (this is the model shown in the Figure 14-6 chart). Not surprisingly, the Figure 14-6 chart shows that Alpina's operations only account for a small percentage of the costs, and that most of the costs are concentrated in Alpina's suppliers. This is a typical situation for most manufacturing organizations, as few companies are fully vertically integrated. What it means to one seeking cost reductions is that the highest returns can probably be found in the supplier base, and not internally.

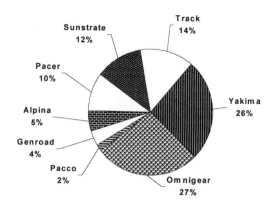

Figure 14-6. Organizational Cost Model for the Alpina Rockchucker Mountain Bike. This model shows that most of the costs are concentrated in Alpina's suppliers.

Under certain circumstances, one might be interested in evaluating only internal organizational costs, as shown in the Figure 14-7 pie chart (for example, a plant manufacturing manager with no supplier responsibilities would want to assess only those costs over which he or she can exercise control). Figure 14-7 shows the extra data associated with Alpina's internal assembly operations, including management, inspection, welding, painting, and integration of all assemblies into completed bicycles.

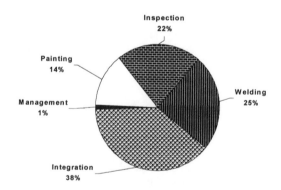

Figure 14-7. Internal Organizational Cost Model for Alpina Bicycles. This model shows the only Alpina's organizational costs for building the Rockchucker mountain bike.

Armed with the information developed in the above cost models, the next step is to act on it by continuing the search for unnecessary costs. This involves targeting the high cost elements and determining their cost makeup. To continue with

148

our Rockchucker mountain bike example, it is apparent from both organizational and component cost modeling perspectives that Omnigear gears are the highest single cost element. Accordingly, it would make sense to work closely with Omnigear to determine if the cost of these gears is reasonable, and in particular, what cost elements associated with the gears can be eliminated or reduced without sacrificing quality.

How does one go about this? In the same systematic manner developed immediately above. Working with Omnigear to develop cost models for the components they supply from component as well as Omnigear supplier perspectives will provide valuable insights into the gearset's cost drivers. These models can be prepared in the same manner as those prepared for Alpina: pie charts based on components and organizations. One can then target the Omnigear component high cost drivers in the same manner as the attack on Alpina's unnecessary costs.

Using Cost Model Information

Once the key cost elements are known (i.e., the high cost drivers identified in the cost models), the challenge is to find the unnecessary costs and eliminate them without affecting the ability of the component to meet its customers expectations. To do this, three sets of data are necessary:

- A listing of each contributor to product cost.

- A definition of customers' expectations (the Quality Function deployment approach, as developed in Chapter 10, is most useful for this).

- An evaluation of each cost element's contribution toward meeting customer expectations.

Developing each contributor to product cost is merely an extension of the cost modeling exercises initiated above. This time, however, the objective is to understand in detail how each cost element contributes to satisfying customer expectations. If a cost element is not required to meet these expectations, it (and the function it supports) should be eliminated.

There are several methods to develop this detailed understanding of a product's costs and how they contribute to satisfying customer expectations. These include business process management, financial data reviews, and again, interviews.

Flowcharting and Business Process Management

The business process management approach is really nothing more than flowcharting and identifying areas in which the process could be improved. This technique is helpful in continuing the cost necessity evaluation. A flow chart showing all of Omnigear's administrative, manufacturing, and inspection operations to produce the Rockchucker gearset will quickly reveal any process steps that add no value. The business process management flow chart shown in Figure 14-8 shows several inspections and a test. The questions are:

- Are all of the redundant inspections necessary?

- Is the test necessary?

- Do any of these process steps add anything to the Omnigear gearset?

If the answer to any of the above questions is no, the step in question should be eliminated.

Organizations are frequently unaware of the level of review, testing, and inspection their products undergo. Reviews, inspections, and tests have a way of creeping into the business process. Typically, they were added long ago to solve specific problems (many of which no longer exist). The problems go away, but the added inspections and tests often linger.

Here's another business process management thought: Take a look at all of your regularly-submitted reports. Frequently (and surprisingly), these are often unnecessary. If the necessity of a report is at all questionable, simply asking the recipient if it is still needed often elicits a cost and labor savings answer.

Financial Data Review

Studying an organization's budgets and other department-level accounting information can be quite revealing. Although the effort can be tedious, questioning the need for every line in a budget spread sheet frequently reveals costs that contribute nothing to the product, and are therefore unnecessary. Manufacturing, engineering, quality, and procurement department budgets are fertile grounds in which to dig for these unnecessary costs.

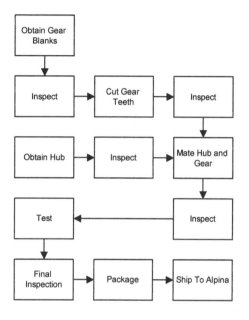

Figure 14-8. A Business Process Management Flow Chart for the Omnigear Rockchucker Gearset. This chart shows several redundant inspections. Are all of them necessary?

Interviews

We covered the need for interviewing personnel involved in the manufacturing operation earlier. The point in our earlier discussion was that those closest to the work often know which manufacturing steps are unnecessary. In a similar manner, those closest to the customer (including the customer) can often provide similar information. The intent here is to determine if previously understood customer requirements and expectations really exist. In Alpina's case, discussions with dealers and bicyclists might reveal if costly Rockchucker features are of no value to the customer. If this is the case, the feature should be eliminated to further reduce unnecessary cost.

Suggested Opportunities

To support the two value improvement approaches discussed above (grabbing the low-hanging fruit and systematically assaulting unnecessary cost), we'd like to suggest several areas to consider in the search for value improvement opportunities. These are summarized in Figure 14-9 and explained in the following paragraphs. Many of the companies with whom we've worked have found enormous cost savings in the areas we'll describe below. Some of the concepts presented below are applicable only to the manufacturing sector, others are applicable only to service industries, and some are worthy of consideration in both areas. All of these concepts can provide a boost to your value improvement brainstorming.

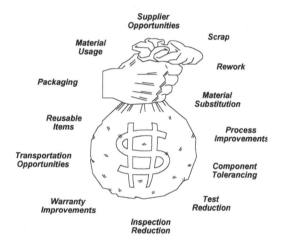

Figure 14-9. Suggested Value Improvement Opportunities. The authors' experience confirms that substantial value improvement can often be realized by investigating the above areas.

Material Usage Opportunities

In many manufacturing processes, material is discarded due to pattern layouts or other process design parameters. For example, if a punch operation is used for cutting slugs out of aluminum ribbon, the pattern layout can be modified as shown in Figure 14-10. The result is a substantial reduction in raw material requirements. Paint spraying techniques can be modified such that lesser amounts of paint are required due to reductions in paint overspray. The question to ask in reviewing any process is: How much material is discarded as a natural consequence of the process, and what can be done to reduce this waste?

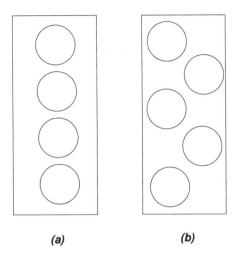

(a) **(b)**

Figure 14-10. An Example of Material Usage Improvement. By rearranging the pattern of slugs cut from an aluminum strip, material usage is decreased 20 percent by going from the layout shown in (a) to that shown in (b).

The answers will lie in two areas. You can implement process or other modifications to reduce the amount of waste, or you can find ways to recycle the waste (e.g., selling the residual aluminum from the first example to a metal recycler). Both approaches should be pursued, but the first usually offers greater value improvement.

Scrap

Evaluating scrap is one of the best ways to find opportunities. The development and analysis of this data are discussed in detail in Chapter 4. By definition, scrap reduction clearly equates to value improvements, and to the extent that scrap reductions can be accomplished economically, they should be pursued. To summarize the discussion in Chapter 4, we recommend tracking scrap costs and rank ordering these costs in a Pareto chart, and then systematically targeting improvements in the highest scrap cost areas.

Rework

Rework reduction is also a superb way to improve value. Again, by definition, eliminating rework

reduces cost. There are many ways to pursue this. One involves pursuing a Pareto-based rank ordering of rework costs as described above and in Chapter 4. Another is to implement a disciplined Cost of Quality System as developed in Chapter 4, and to use this information to target high rework areas for process and other improvements. The Tennant Company (manufacturers of floor sweeping equipment) took a radical approach and simply eliminated rework areas. Tennant found that simply eliminating the rework function (i.e., reassigning workers who performed nothing but rework, thereby doing away with the rework stations) instilled in their workers a recognition of the need to build Tennant floor sweepers right the first time. Tennant's action simultaneously eliminated rework and improved the quality of their products.

Packaging

In many instances, packaging can either be improved or simplified (or both) to reduce cost. Suppose one learns that nonconformances are frequently induced by material handling damage. Corrective action in this situation could include packaging improvements to protect the product from such damage. Packaging improvements will add cost, but the added cost will result in greatly reduced scrap or rework on the items being damaged. In other cases, packaging designs may be well in excess of what is really needed to protect the product.

We remember a company buying small fasteners for use on a metal canister. The fastener supplier boxed each fastener individually, with protective foam padding in each box. The fasteners cost less than $0.18 apiece. The packaging brought the price to about $0.40. Manufacturing engineers in the company buying the fasteners found that the extra packaging wasn't necessary, and that time was lost on the assembly line when the assemblers had to remove each fastener from its container. The workers recognized this, too, and they recognized it long before the manufacturing engineers discovered the problem. The assembly line people assigned a person to take all of the fasteners out of the individual boxes and dump them in a single large box. This allowed the assembly line workers to simply take what they needed when they needed it.

The above practice had been going on for nearly 2 years when someone recognized that the company was paying for unnecessary packaging. When the canister manufacturer contacted the fastener supplier, the company learned that they were the fastener supplier's only customer that had specified the individual packages. No one at either company knew why. Here's what resulted when the requirement for individual packaging was dropped:

- Both companies monitored new shipments to assure no handling damage resulted (none did).

- The extra position at the canister manufacturer was eliminated (i.e., the person who took the fasteners out of their individual boxes and put them in the single large box).

- The cost of the fasteners dropped to less than half of the original cost.

The above is a particularly good value improvement example, as it illustrates the utility of several concepts previously discussed: business process management, quality function deployment, interviews, employee involvement, and others.

Reusable Items

We recommend taking a hard look at anything that is discarded during the production or service process, and determining if there is a way to reuse it. Packaging and assembly aids frequently offer such opportunities.

We once worked with a company that used a metal device as an assembly aid. The company built its product at a facility in California, installed the assembly aid, and then shipped both to another facility in Kansas (where it was incorporated into the final assembly). In Kansas, the metal assembly aid was discarded when the subassembly was installed into the final assembly.

The company in California built more than four million of these subassemblies every year. Each one required an assembly aid that cost 25 cents. As the result of an employee suggestion, the assembly aids were collected at the Kansas facility, and instead of being thrown away after one use, they were returned to the California facility for reuse.

The California company achieved savings in excess of one million dollars the first year. As this example demonstrates, reuse can offer powerful value improvement. Other reuse opportunities may be found in packaging, bottling, or anything else normally discarded as part of the production, delivery sequence, or life of a product.

Material Substitution

In many cases, lower cost materials can be substituted to improve value. This is an area best left to a company's engineering department, and careful attention should be paid to strength versus load, corrosion resistance, weight, and other design considerations. We've seen many instances in which companies realized significant cost reductions without sacrificing product quality by substituting plastics for metals, carbon steels for stainless steels, etc.

Process Opportunities

Perhaps no other area offers greater value improvement opportunity than does process improvement. Here are several areas in which we've been able to realize value improvements:

- *Bottleneck Elimination.* One quick way to find unnecessary cost is to simply wander around a factory or office and look for bottlenecks. The concept here is that anytime the production flow is interrupted (as occurs with bottlenecks), the process induces unnecessary cost (due to idle time, unused inventories, etc.). Anything done to eliminate the bottlenecks, therefore, automatically reduces costs without sacrificing the satisfaction of customer expectations. We recommend Goldratt's and Cox's excellent book, *The Goal*, which further develops this concept.

- *Task Reallocation.* Most manufacturing companies procure many of the components and subassemblies in their products from suppliers. Service organizations often do the same. This is often done in an undisciplined manner, without regard for where the work might be done most economically. Sometimes it may appear that a supplier can do the work for less, but often quality deficiencies, subsequent price increases, or other factors can reverse the desirability of a make versus buy decision that a company may have made years ago. Even if it still makes sense to buy the item, we have often discovered additional savings by examining everything the supplier does to the product, and evaluating if these individual actions can be accomplished for less by the company buying the service or supply. Sometimes a subcontracted item requires additional processing once the item is delivered to the buyer (for example, your company may apply markings or other identifying features to subcontracted items). Having the supplier apply the markings will usually cost less than what it will cost your company to do the work (this is because as one works down the supplier chain, lower overheads are usually the case). The message here is also simple: If your company buys goods or services, examine where the work can be done for the least cost, and consider all factors (overhead, current labor and material costs, cost of defective goods, transportation costs, etc.).

- *Automation*. Opportunities for automation exist in both manufacturing and service organizations wherever operations are performed manually. A treatise on modern manufacturing automation methods is beyond the scope of this book, but one searching for value improvement in a production process should consider identifying all manufacturing operations being performed manually, and requesting the company's manufacturing engineering department to evaluate the feasibility of automating the operation. Similarly, numerous excellent software packages are available to automate manually-performed service operations (examples include using a desktop publishing approach to replace cutting and pasting illustrations in text, or using electronic spreadsheets to replace routine numerical operations performed manually with calculators or adding machines).

- *Labor Reduction*. Organizations tend to add staff during prosperous times, often without carefully evaluating the need for additional personnel. Although staff reductions are less preferable to us as a value improvement measure (simply because organizations often

jump to layoffs as a panacea for cost reduction, without taking advantage of the opportunities discussed in this chapter), the simple fact is that savings can frequently be attained by selectively and intelligently pruning the workforce. Again, a business process management approach is recommended. Flowchart and understand the process, evaluate the functions of each position against the needs of the process, and then eliminate redundancies or non-value-adding positions.

- *Assembly Aid Incorporation*. As one evaluates manufacturing processes for value improvement, excessive scrap, rework, setup times, and bottlenecks often indicate opportunities. The use of assembly aids (special tools to hold the product, allow easier assembly, guide parts together, etc.) can frequently reduce cost by eliminating or controlling the conditions that induce the cost drivers. When such areas are discovered, we recommend turning to the manufacturing engineering department for assistance.

- *Material and Paperwork Movement Reductions*. Business process management can also help by identifying excessive material movement. When flowcharting manufacturing or administrative processes, pay particular attention to paperwork and material movement. Reducing material and paperwork movement will improve value two ways: It will immediately reduce labor and other costs associated with the movement, and it will streamline the process, thereby providing ancillary value improvement gains. Although many companies ignore this opportunity, this is not an area to be overlooked. Excessive material and paperwork movement is very common. The cost reduction and other benefits to be realized from such improvements are often significant.

- *Statistical Process Control Implementation*. Chapter 10 described statistical process control and its usefulness as a value improvement tool. We advise our readers to consider statistical process control wherever high scrap and rework are present. Statistical process control offers huge potential for reducing cost (by reducing and often eliminating scrap and rework) and for improving value (as a consequence of both cost and variability reduction).

Component Tolerancing

Several areas involving value improvement through product redesign have already been discussed (material substitution, redesign of the product to simplify assembly, etc.). Component tolerancing is another design factor that should be carefully examined for value improvement opportunities.

A dimensional tolerance is the degree of control one must exercise over a part's dimensions. Component dimensions on manufacturing drawings are never exact. For example, if a part needs to be one inch long, engineers will always specify the basic dimension (i.e., the one inch dimension) with a tolerance. The fully dimensioned part might carry a dimension such as 1.00 ± 0.030, which means that the part is acceptable if its length lies anywhere between 0.970 to 1.030 inches. Engineers do this because they (and the machines and machinists that must make the parts) understand that they can never make parts that are exactly one inch in length, so the tolerance is needed to tell everyone how much "tolerance" (hence the name) is acceptable in manufacturing the part.

Careless tolerancing often adds unnecessary cost to a product. This occurs because as tolerances are tightened, machining costs generally increase significantly (especially if the part is not being manufactured in a statistically controlled process, as explained in Chapter 10). More time and care are needed to produce a part that lies between 0.970 and 1.030 inches than is needed to produce a part that lies between, say, 0.50 and 1.50 inches. The dimensional difference need not be that dramatic, though, to produce a significant difference in manufacturing cost. Increasing tolerance just a few thousandths of an inch can frequently lower manufacturing costs significantly.

The people who assign component tolerances are frequently draftsmen who do not understand manufacturing processes and the tolerances they are capable of meeting, or the needs of the design to assure reliable function and component interchangeability. The result is that most draftsmen err on the conservative side, and assign tolerances that are unnecessarily stringent. Many draftsmen mistakenly believe that assigning tight tolerances somehow results in higher quality, when in fact, all it does is drive up scrap, rework, and other manufacturing costs.

The result of the above is that sizable value improvements can often be realized by examining all areas of high scrap and rework, and areas where high manufacturing time is required. If these conditions are induced by attempting to meet a tight tolerance, one simply needs to determine if the tolerance can be relaxed.

A word of caution is in order, though. Most engineers will naturally defend tolerances on drawings they created (even though draftsmen create the drawings, engineers typically sign them). The best approach is to approach the engineer in a nonjudgmental manner, and ask how tight the tolerance has to be. Given this approach, many engineers are proud of their abilities to show that relaxing a tight tolerance is acceptable.

Transportation Opportunities

Travel is inherent to most businesses, and the travel budgets for most companies are sizable (it's not unusual for a company with sales in the $50 to $100 million range to have a travel budget of a million dollars or more). With only a moderate amount of planning, these costs can often be cut in half. One approach is to simply do more business by phone. It's probably true that face-to-face contact promotes better business, but think back on recent business trips, and ask yourself if many of these could have been accomplished by phone.

Another cost reduction opportunity involves questioning the need for travel. We have found that asking prospective travelers to list the objectives of planned trips and why each person needs to travel (if more than one person is going) will frequently result in travel expense reduction.

Planning trips more than a day or two in advance can result in sharp air fare reductions. Most airlines' rates increase sharply for short-notice reservations, so making reservations early can offer a good payback. This involves a measure of planning discipline, but the rewards are significant. If your company uses a travel agency, make the agency aware of your desires to control costs. Travel agencies earn fees based on airline fees, so it is not usually to their financial advantage to find the least expensive option (unless, of course, the agency is made to understand that your business will go elsewhere if they fail to support your value improvement initiatives in this area).

First class air travel is another enormous value improvement opportunity. Some companies allow their senior executives to fly first class. We've never been able to understand or justify the exorbitant rates charged for first class air travel, and we cannot help but believe that most people feel the same way. If you want to gain an insight into just how true this is, ask any executive who travels first class on business (when his or her employer is paying for the travel) if they travel first class on personal business or vacations.

Inspection and Test Opportunities

Many tests and inspections are redundant. We've often seen instances where identical tests are performed on the same item at different points in the process with no fallout occurring over years of testing. The same is often true for inspection points. Statistical process control implementation will frequently eliminate the need for inspection and test, but even if your company opts not to pursue statistical process control implementation, value improvements can often be realized by eliminating redundant tests and inspections. The best way to find these is to turn to business process management and flowcharting, as mentioned earlier. Flowcharting the entire process and identifying all tests and inspections will frequently identify redundancies, which immediately become candidates for elimination.

We also recommend examining the reject rate at each inspection and test point. Even if the test or inspection is not redundant, if there are very few or no failures it may make sense to inspect or test the item less frequently. If there are no failures or rejects, it may make sense to eliminate the inspection or test altogether.

Warranty Opportunities

Nearly all commercial organizations (especially those involved in producing retail products) and many defense contractors warrant their products. Warranty costs generally fall into three categories:

- The costs associated with administering the product.

- The costs associated with establishing a financial reserve for potential warranty claims.

- The actual costs associated with replacing defective product.

Warranty administration costs are no different than any other administrative cost, and these costs can be evaluated using the techniques described earlier in this chapter. In our experience, companies tend to err on the high side in establishing warranty reserve to offset future potential claims. The subject is far too specialized to treat adequately here, but our recommendation is to question the basis for any warranty reserve (simply asking how a warranty reserve was established will often result in reductions).

Manufacturers can lower costs and provide better warranty service by building and storing extra product (instead of building replacement product only when warranty claims are exercised). These costs, however, must be balanced against the costs of carrying the increased inventory.

Supplier Opportunities

As explained earlier, most manufacturing companies and many service companies subcontract significant portions of their products and services. Consequently, it makes sense that significant opportunities for value improvement exist in an organization's supplier community. In addition to staffing your organization's procurement department with the best people you can find (and in particular, with skilled negotiators), there are numerous other approaches to reducing procured item costs.

The first of these opportunities lies in the concept of multiple buys. In many cases, increasing the order quantity will result in lowered prices. The concept is to find uses for the additional quantity of a purchased item in other areas (such that a larger purchase quantity results). If you can procure items for more than a single product or program, it may make sense to do so if the cost advantages outweigh the advantages of carrying additional inventories or work in process. This can be done across program or product lines, and in certain cases, across facility lines.

One company we worked with used an epoxy that it ordered in small quantities (at considerable expense). This company learned that another division of its parent corporation used the same epoxy in much larger quantities. The smaller company combined its purchases with those of its sister division for a sizable cost reduction. The company also achieved a significant quality improvement. Formerly, its smaller orders were created on demand, and significant variability in epoxy bond strength was a persistent problem. When the smaller organization's purchase orders were combined with the larger division's, the epoxy company created larger batches of product for both companies, with much less product variability. The results included improved bond strength and consistency.

Multiyear buys are a variation on the multiple buy concept. This approach works particularly well in government contracting, where orders tend to be placed for one year's worth of services or products. If you can convince your customer to contract for more than a single year, the added several years of services or products can be flowed to your suppliers. This practice often results in significant price reductions. Another company we worked with managed to reduce the price of its products more than 90 percent over a dozen years through multi-year contracting arrangements, while simultaneously and steadily increasing its profits! Even if you cannot obtain the assurances of a multiyear contract, you might consider offering a multiyear contract to selected suppliers. The cost advantages could outweigh the added risk.

Yet another variation of the multiple buy concept is to team with selected suppliers through an exclusive arrangement. Many companies use this approach, which offers the potential of both reduced cost as well as improved quality. The reasons for the cost reductions are obvious (larger quantities, decreased supplier marketing costs, etc.). The quality improvements result from steadier supplier production operations (which result in less product variability). Exclusive supplier arrangements also result in suppliers that better understand the needs of their customers (with this improved knowledge of customer requirements, improved quality naturally results). Many companies offer exclusive supplier relationships only to those suppliers meeting rigid quality requirements (statistical process control, above-average product reliability, in-place quality measurement systems, etc.). Again, this results in significantly improved supplier quality, with a simultaneous cost reduction.

No discussion on supplier cost reduction would be complete without mentioning competition. Competition probably represents the greatest supplier cost reduction opportunity. A word of caution is in order, though. Mindless competition focused exclusively on cost is generally accepted as detrimental to continuous quality improvement. If cost-based competition is pursued in lieu of the supplier teaming concepts discussed above, our recommendation is simple: Remember that price is but one element of cost, and that other elements include poor quality, schedule slips, and other factors that may well wipe out any cost gains resulting from a simple cost competition.

Overcoming Value Improvement Obstacles

Finding and implementing value improvements are two very different things. In the enthusiasm and excitement of developing value improvements, it's easy to lose sight of the fact that not everyone will agree with the obvious logic inherent to the discovery and implementation of value improvements. There are many reasons for this. There's the natural resistance to change most of us share. Some people will be threatened by readily apparent cost reductions, perhaps fearing retribution for not having discovered and implemented the change earlier. Suppliers will see threats to profits and sales. There may be other, more subtle constraints. This section explores a few of the more common obstacles, and how to overcome them.

Implementation Issues

There are two issues associated with resistance to value improvement implementations: cost and risk. There are numerous methods available for evaluating the investment required (particularly when compared to implementation costs). These financial analysis tools include internal rate of return analysis, net present value analysis, payback period analysis, and others. All are beyond the scope of this book, but you should be aware that they exist, and seek assistance from your organization's financial analysis group to assist in determining if the value improvement concept has merit. The other issue, risk, has to do with what might occur when value improvement concepts are implemented. The approach for managing value improvement risk is to identify all potential outcomes, and what has to be done to control any unacceptable outcomes.

Internal Resistance

Internal resistance is perhaps the most difficult obstacle to overcome when developing and implementing value improvements. There are several potential reasons for such resistance (including many that may be legitimate). When the resistance does not seem to have a sound basis (either financially or technically), the results probably lie elsewhere. Our experience indicates that in many instances, simple resistance to change may be the culprit. Fear is also a factor here. We suspect that managers may harbor fears that they will somehow be blamed for not recognizing the cost reduction and implementing it sooner (the logical response to such a fear is to deny the validity of the improvement). In other cases, the objection may simply be an outgrowth of the "not invented here" syndrome. The good news is that all of these objections can be overcome through classic change management skills. In particular, the ability to instill ownership of the suggestion in the person objecting to the change is quite helpful. Convincing someone who objects to a value improvement that the idea was really his or hers is particularly valuable.

Supplier Reluctance

Here's a situation that frequently develops when working cost reduction issues with suppliers: The supplier turns a cold shoulder to such efforts. The

reasons are straightforward. Many suppliers see cost reduction as a direct threat to both sales and profits. The value improvement responses we've found to be successful in such situations includes:

- *Logic*. Frequently, explaining to suppliers that reduced costs resulting in improved value inevitably lead to both profit and sales growth. Enlightened suppliers will usually accept this logic.

- *Competition*. Very few organizations enjoy a lock on the products or services they provide. Virtually all intelligent suppliers will become remarkably cooperative if they are convinced that customers will move to a competitor to secure value improvement cooperation.

- *Sharing Arrangements*. In certain situations (e.g., when dealing with sole source suppliers) it makes sense to offer value improvement sharing arrangements with suppliers. We've been involved in arrangements in which the savings resulting from supplier-implemented cost reductions were shared with the supplier, instead of being realized only by the customer. Suppose, for example, that you discover a way to reduce one of your supplier's costs by 20 percent. The supplier doesn't have to implement the change, and he certainly doesn't want to give of 20 percent of his sales base (and the corresponding profit associated with it). Instead of you pushing the supplier to do so, suppose you offer to split the savings such that the supplier eliminates the 20 percent of added cost, but gets to keep 10 percent of the savings (which equates to a 10 percent profit increase). You only attain a 10 percent cost reduction (but 10 percent is better than nothing), and in the process of lowering your costs, you gain a more profitable supplier.

Summary

This chapter began with a discussion on the meaning of value improvement. When taken beyond simple cost reduction, value improvement is strongly linked to quality improvement. Many value improvement ideas can be developed through employee involvement approaches that include brainstorming and suggestion programs. Other value improvements are often available through a more disciplined approach that involves a systematic assault on unnecessary costs.

References

Commonsense Manufacturing Management, John S. Rydz, Harper and Row, 1990.

Value Analysis, Carlos Fallon, John Wiley and Sons, 1980.

Techniques of Value Analysis and Engineering, Lawrence D. Miles, McGraw-Hill Book Company, 1972.

"Value Engineering and Complex Munition Systems," Joseph Berk and Jean Starnes, *Proceedings of the 1989 Society of American Value Engineers International Conference*, 1989.

Well Made in America, Peter C. Reid, McGraw-Hill Publishing Company, 1990.

Chapter 15

Supplier Teaming and Procurement Quality Assurance

Including a key ingredient in the quality management formula...

Alex Johnson reviewed Bernard Medical Devices' weekly quality summary with disappointment. When Johnson took over as general manager of Bernard Medical Devices last year, he felt confident that the company would be able to lower its number of inprocess rejections. The blood chemistry analyzers Bernard Medical manufactured were complex, but Johnson felt if he continued to emphasize the importance of continuous improvement and provide the right training to his staff, the defect rates would have to come down.

At first, the nonconformance rate decreased in accordance with Johnson's expectations. But after the first six months, it remained stable. There were typically about 40 nonconforming material reports generated each week. Sometimes the number was a little higher, and sometimes a little lower, but the big breakthrough Johnson wanted had just not materialized. Johnson looked at the report in front of him again. It showed 41 nonconformances for the preceding week, most of which were detected in the assembly area during system checkout.

Alex Johnson picked up his telephone and called Kevin Andreason (Bernard's director of manufacturing), Buck Forgia (the director of quality assurance), and Tony Pierson (the assembly area supervisor). He asked all of them to come to his office, and in a few minutes, all of the men were in front of him.

"Anybody want coffee?" Johnson asked. Forgia said yes and helped himself to a cup from the brewer behind Johnson's desk. The other two men declined. "I'd like to get back to a subject we've been addressing for the last year," Johnson said, "and that's the nonconformance rate. I know you guys are working this issue hard, but in the last several months we just haven't made any significant progress. We still come in with around 40 reject reports every week. The quality summaries Buck's guys issue show that it appears to be focused in the assembly area, Tony, and that's why I wanted you in this meeting. I'm looking for help. What do you guys suggest?"

"Alex, you're absolutely right, most of the nonconformances are being detected in my area," Tony Pierson said. "We don't have any assembly workmanship problems, though. We put the components and subassemblies we are given together, and sometimes they just don't work. The problem is with the subassemblies we're given, not with our workmanship."

"Is that the bulk of it? Most of the problems are due to subassembly or component deficiencies?" Johnson asked.

"That's true," Andreason and Forgia both answered. "We get a lot of problems with purchased items," Andreason continued. "They get through Receiving Inspection, but then they conk out in the assembly area."

"Why is that?" Johnson asked Forgia.

"Mostly, it's because we can't inspect for everything in Receiving Inspection," Forgia said, "there are performance requirements our suppliers don't meet."

"Hmmm," Johnson answered, lost in thought. "Have either of you talked to Parton?"

Nat Parton headed up Bernard Medical Devices' procurement department. Andreason and Forgia both shook their heads to indicate they had not discussed the problem with Parton. Johnson picked up his phone, in a few minutes Parton joined the meeting. Parton helped himself to the coffee without waiting to be asked.

"Nat, are you aware of the problems we've been having with procured components in the assembly area?" Johnson asked.

"Yes, I am," Parton answered. "We get an assortment of, oh, 25 or 30 items every week that fail and have to be returned to the suppliers we buy them from."

Johnson sat up in his chair. He felt he was really on to something, perhaps the quality breakthrough for which he had been searching. "What do the suppliers say when we return these things?" he asked.

"Not much," Parton answered. "They usually run them through their acceptance tests again, and they usually pass. They bill us for the test, and then send the parts back to us. Sometimes they work the second time they come back, sometimes they don't."

"Has anyone from engineering looked at the acceptance tests the suppliers perform, or how we test them in Receiving Inspection?" Johnson asked. The room fell silent.

Many people lose sight of the fact that even if a quality improvement program could make everything in a manufacturing facility perfect, more than half the work would still be ignored. The missing key ingredient that many of us fail to initially consider is the supplier base, those companies that sell goods and services to us.

Why is this so critically important? Most manufacturing companies spend more than half of their sales dollars on purchased parts and services (see Figure 15-1). Honda of America recently stated that more than 80 percent of their costs are due to suppliers of goods and services. About 65 percent of the Tennant Company's costs go to suppliers. For defense contractors, the number seems to hover around 70 to 80 percent. We recall one program, an

aerially delivered mine system used with great success in the Persian Gulf war, that was literally 100 percent subcontracted. Gencorp, the corporation holding the prime contract, purchased all of the components from external suppliers, and then paid another supplier to assemble them into complete mine systems.

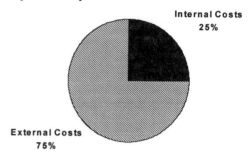

Figure 15-1. Typical Supplier Content. Most companies spend more than half their sales dollars on supplier goods and services. Extending the quality management process to an organization's suppliers makes good business sense.

What does all of this mean to an organization that wants to improve quality? The typically high supplier content of all goods and services won't allow an organization to ignore suppliers if an effective quality improvement process is to be implemented. Suppliers simply form too big a piece of the pie. Organizations have to recognize that in order to improve quality, they also have to help their suppliers be successful.

Is helping suppliers improve quality a realistic goal? Don't organizations have enough to do just improving quality internally? Ford Motor Company and IBM take the challenge so seriously that they advertise their success in helping their suppliers attain world class product quality. Both organizations placed full page advertisements in *The Wall Street Journal* and *USA Today*. The Ford and IBM advertisements were not focused on Ford or IBM quality, but instead on the quality of selected Ford and IBM suppliers. Ford and IBM sent powerful messages with these advertisements. All Ford and IBM suppliers (and their potential suppliers) immediately recognized that Ford and IBM are serious about supplier quality. The featured suppliers (and their potential customers) recognized that Ford and IBM are serious about supplier quality. Most importantly, Ford and IBM customers (and potential customers) realized that Ford and IBM are serious about supplier quality, and by an extension of this logic, serious about Ford and IBM final product quality.

What does all this mean to an organization managing quality in a high technology environment? Simply that suppliers have to be included in the process. How does one go about doing this? We'd like to suggest a number of considerations, which are explained in the following paragraphs and illustrated in Figure 15-2. Our research and experience have convinced us that the following concepts are key to successfully extending the quality management process to suppliers:

- Effectively communicating with suppliers.

- Making suppliers part of the development process (for both product and process design), and part of the continuous improvement process.

- Focusing on suppliers' cost structures, rather than suppliers' prices.

- Emphasizing long term relationships with fewer suppliers.

- Objectively evaluating supplier performance, and selecting suppliers based on the results of these evaluations.

Let's explore each of the above subjects in greater detail, and how to extend the quality management process to an organization's suppliers.

Communicating Effectively With Suppliers

Communication is a basic part of quality management, and that holds true for suppliers as well. Philip Crosby suggests that perhaps one of the best facts to communicate to suppliers is that the buying organization expects all of its requirements to be met. That may seem so obvious as to almost not be worthy of mention, but our experience tracks with Crosby's recommendation. We've frequently encountered instances in which suppliers seemed to be unaware of their customers' intent to abide by the purchase order requirements. Simply explaining to

suppliers that they will be held accountable for meeting purchase order requirements helps to focus their attention.

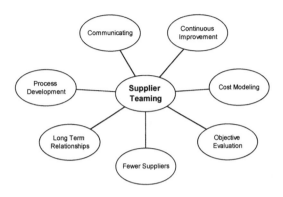

Figure 15-2. Keys to Extending Quality Improvement. Each of the above concepts is critical to successfully extending TQM implementation to the supplier base.

We've found it helpful, as a customer, to explain to suppliers that customers are not their inspection departments. On occasion suppliers get sloppy in their final inspection operations, reasoning that if they ship defective product it will be detected during the customer's receiving inspection. Suppliers that rely on customers for detecting their defects don't deserve future business, and customers need to let suppliers know that.

Crosby makes a point of recommending that customers let suppliers know when their product fails to meet requirements, and again, our experience tracks closely with Crosby's. Although it seems incredible, customers frequently scrap or rework supplier hardware without letting the supplier know. Under these circumstances, a supplier could never be expected to improve. Sometimes this situation develops because the customer feels it's more expeditious to simply scrap or rework the hardware in house. Sometimes it occurs because the right people in the customer organization don't know it's going on. Whatever the reason, it only makes good business sense to inform suppliers when their hardware or services do not meet requirements. Crosby recommends going right to the top when this occurs, and calling the supplier's president.

Many organizations have supplier quality forums, in which key suppliers are invited to the customer facility to review the customer's quality requirements and continuous improvement objectives. These meetings can be particularly valuable for explaining customer quality improvement initiatives and objectives, and inviting supplier participation. They also offer an excellent opportunity for educating suppliers on customer quality strategies, supplier quality problem areas, and general quality improvement concepts. Many organizations (Harley-Davidson, for one), even go to their suppliers' facilities to conduct in-house training on just-in-time inventory management, statistical methods, employee involvement, teams, and other quality improvement practices. Periodic quality improvement seminars with suppliers also provide opportunities for suppliers to tell customers what they could do to improve quality (some companies, such as Black and Decker, even ask their suppliers to rate them as customers).

There's one other supplier communications concept worth mentioning, and that's communications control. We encourage everyone in customer organizations to speak frequently with their supplier counterparts, but everyone needs to understand that direction to a supplier needs to be funneled through the customer's procurement organization. Customers and suppliers can get into trouble fast if engineers, quality engineers, or others outside of the procurement organization start giving direction to a supplier and the supplier accepts the direction.

Making Suppliers Part of the Solution

As explained earlier, purchased parts and services make up at least half of the product in almost every industry. It seems intuitive that at least half of all product development efforts, quality problem resolutions, and continuous improvement projects would be influenced by supplier components or services. Our intuition notwithstanding, we've frequently observed instances in which companies attempt product development efforts, quality problem resolutions, and continuous improvement projects without involving suppliers. In fact, our observations indicate this seems to be the case more often than not.

We strongly recommend including suppliers during development programs, when attacking problems, and when attempting to formulate and meet continuous improvement objectives. Successful companies integrate the activities of their internal organizations with their suppliers' corresponding departments for this purpose. McDonnell Douglas Electronic Systems Company engineers and quality engineers work closely with their counterparts on new development programs and upgrades for aircraft electronic systems. Honda of America has stated that it couldn't meet its short product development times without including suppliers at the outset of a development effort.

Why establish such a close link with suppliers? If the problem is associated with a supplier component, or if quality improvements are desired in a supplier component, the supplier can offer valuable insights. Integrating an organization's engineering, manufacturing, and quality teams with their supplier counterparts on a project-specific basis assures the incorporation of supplier inputs early in the process, and this will prevent pursuing approaches that the suppliers know won't work. It will also result in suppliers suggesting approaches on their equipment that they know will work, allowing a customer to take advantage of supplier expertise.

Does the above process work? Consider the relationship between Intel and most of the IBM-compatible computer manufacturers. As soon as Intel is ready to market a new microprocessor, the computer manufacturers have designs based on the new chips ready to go into production. That doesn't happen as a result of the computer companies waiting for Intel to complete development of its microprocessors and then release design information. The computer manufacturers don't design systems and then write a microprocessor specification for Intel. At each new Intel introduction for the 286, 386, and 486, and Pentium series of chips, the computer manufacturers had system designs based on the new microprocessors in production. The computer manufacturers work closely with their supplier, Intel, and highly reliable products emerge after short development cycles. The concept, which is also being used successfully by automobile manufacturers and others, is applicable to any industry.

Focusing on Costs Rather Than Prices

Businesses in America frequently compete their purchases among suppliers, with cost as the sole determinant of success. Competition is inherent to our culture. Competition can be a good thing, but in our country's history the basis for supplier competition evolved into one based solely on price. We systematically award business to the low bidder. Quality always plays a role, as does technology, but in industry today, price most frequently dominates procurement decisions.

Many businesses compound the problem by repeating the competitive bidding process annually, or more often if the opportunity arises. Companies award a program to the low bidder, a relationship starts to develop, and then another supplier underbids the first one and steals the business away (frequently with no consideration of the technical risk, schedule risk, or quality risk).

Does this really happen? Absolutely. We've seen instances in which customers let suppliers take business away from each other for less than a penny a part. What can this do to the customer? In one case, when the newer and cheaper supplier couldn't perform, it shut down a two million dollar a week production line for six weeks until the first supplier could retool to resume production. The message here is that supplier prices certainly play a role, but not the only role, and not necessarily the lead role. Customers need to focus on the total cost of doing business with a supplier, and not worry simply about the initial procurement price.

Supplier Cost Modeling

Polaroid uses supplier cost modeling to help their business keep supplier prices down while simultaneously allowing them to do business with quality suppliers. A cost model represents the costs suppliers incur to manufacture the product the customer buys. The cost modeling concept discussed here is the same as discussed in the previous chapter on value improvement.

How does one go about constructing a supplier cost model? Cost information has to be obtained gradually (most suppliers will not simply give it to a customer). Once this is done, though, the customer

may have a better understanding of a supplier's costs than the supplier does. Polaroid operates on the belief that a supplier is entitled to a fair profit, and they help their buyers obtain this by helping them to see where their true costs are (and what can be done to control these costs). Harley-Davidson operates similarly, often sending teams to suppliers to help them simultaneously improve quality while reducing costs.

Longer Term Relationships, Fewer Suppliers

We addressed the fallacies inherent to constantly changing suppliers to lower cost. Switching from supplier to supplier simply to save a few dollars doesn't make sense from a quality perspective, and in fact, it often ends up costing more to do so. Managers will argue that using a smaller number of suppliers or just a single supplier will raise costs. Philip Crosby puts the problem in perspective by asking his students to think about the people they do business with in their personal lives. Most of us have one dentist, one barber, and one of each of the people we rely on. Would you consider changing dentists or barbers to save money? Most of us would not, because we usually develop stable business relationships with people who provide important services to us, we feel good about the service we get, we would not want to deal with more than one in each category, and we would not want to start over again with a new dentist or a new barber. We also feel comfortable with the cost of the services they provide. Perhaps we could do better from a cost perspective, but we intuitively realize it probably wouldn't be worth the bother, and the risks might be unacceptable.

The same logic applies to working with suppliers. If your organization has a good relationship with a quality supplier, does it make sense to switch to another supplier to save a few dollars? Usually, the answer is no, but that doesn't stop most of American industry from doing so on a routine basis.

Long term agreements with fewer suppliers, based on more than just supplier price, offer numerous advantages to both the customer and the supplier. Long term agreements allow suppliers to invest in capital equipment, which can favorably affect quality and cost. Suppliers can buy raw materials at better prices if they know they have a long term program. Suppliers are often more willing to follow a customer's quality management lead on statistical process control and other initiatives if the arrangement is for more than a single procurement. Working with a smaller number of suppliers also serves to reduce variability. Procuring a component or service from one supplier inherently offers less variability than would procuring the same component or service from two, or three, or more suppliers. Working a long term agreement with fewer suppliers also helps to support a customer's inventory reduction process, because the suppliers can better learn their customer's schedule needs, and time their deliveries to meet the customer's needs.

What factors help companies determine which suppliers are candidates for long-term business relationships? There are a number of factors in this decision, but certainly a willingness to work with customers on controlling costs (as outlined above), product quality, schedule performance, supplier surveys, and past performance figure prominently. Sometimes technology plays a key role. Quality always enters the equation though, as it is perhaps the strongest indicator a customer has of the true cost of doing business with a supplier.

Harley-Davidson emphasizes two to three year agreements with their suppliers. Most of the automobile manufacturers in this country are doing the same. Other companies are similarly pursuing this goal. The defense industry has been somewhat slower to respond, primarily because their customer, the U.S. government, often awards contracts on a yearly basis.

In moving to long term commitments with suppliers, most companies are also reducing the number of suppliers they do business with. The concept is to get down to a few truly outstanding suppliers, and continue to do business with these. The perceived advantage in doing business with a multitude of suppliers is price competition. In many cases competition hasn't served that goal well (if the total price of doing business with a supplier is considered).

This concept of reducing the supplier base is gaining momentum. The reductions in many companies' supplier bases are quite startling. The Yale Lock company went from 1,100 suppliers to 650, and their goal is to get down to 150. Loral Aeronutronic, a high technology defense contractor, has gone from 5,000 to 1,000 suppliers. Other companies are pursuing similar reductions.

Sometimes both suppliers and customers shy away from long term agreements because they fear fluctuations in raw material costs or other factors are beyond their control, or they feel the customer's business base is uncertain. Price fluctuations can be structured into the long term business agreement, with both parties agreeing to renegotiate if raw material, labor, or perhaps other supplier costs go up or down by more than a specified percentage. Customers can address fluctuations in their business base in long term supplier agreements by offering to give the supplier a specified percentage of the customer's business base, rather than a fixed dollar amount.

Evaluating Suppliers

As mentioned earlier, evaluating suppliers with the intent to reduce the number an organization does business with requires considering several factors. These factors include delivery performance, quality performance, cost, and perhaps, technology or other factors. In Chapter 4, on quality measurement systems, we developed a concept for rating suppliers based on their quality performance, so the subject won't be covered again here. We have encountered several supplier rating concepts that work well, though, and these will be discussed.

Multi-layer supplier ratings and the certified supplier concept go beyond the information on supplier rating systems presented earlier, and the success some companies have attained using these approaches bears mentioning. James River is an organization that grades its suppliers as approved, qualified, preferred, and certified. James River uses the concept as a quality ladder for suppliers to climb, with additional advantages offered to suppliers as they climb from merely being approved to being certified. The concept is that as suppliers' quality performance improves, they face less stringent inspection requirements and are offered other benefits.

Some companies that certify their suppliers perform no receiving inspection at all. Material from certified suppliers goes directly from the receiving dock into the stock room (or better yet, directly to the production line), with no delays for receiving inspection. The concept is that a certified supplier has proven their quality system prevents defective material from reaching the customer, and therefore, no additional inspection is required.

Sargent-Fletcher, another defense contractor, certifies a few of its suppliers, recognizes the cost advantages these suppliers' superior quality offers, and passes some of the financial advantage back to the supplier. Once a supplier has been certified by Sargent-Fletcher, the supplier's shipments are no longer subjected to receiving inspection, and the supplier is offered a pricing advantage. Sargent-Fletcher still seeks bids from other suppliers, but they have to have at least 10 percent lower costs before they will be considered as a certified supplier replacement.

IBM takes a slightly different tack, and negotiates quality incentives into some of its supplier contracts. As defect rates decrease, premiums are offered to IBM cable suppliers. The bonus for superior quality can be significant, and IBM has seen its cable suppliers achieve significant quality improvements as a result.

Supplier Surveys

Perhaps the oldest means of attempting to evaluate supplier performance is the supplier survey. This practice consists of a customer team visiting the supplier's facility to observe its facilities, equipment, work force, quality assurance and engineering organizations, quality assurance procedures, calibration system, and other items considered relevant to its quality. Periodic quality surveys are a requirement for certain types of complex equipment suppliers in the defense industry (government procuring agencies are required to survey prime contractors, and prime contractors are required to survey their suppliers).

Based on our experience, supplier surveys are necessary from a government requirements perspective in the defense industry, and we believe face-to-face meetings in supplier facilities are useful for good communications, but the real proof of a supplier's ability to deliver quality hardware is the

supplier's track record. The bottom line is that one should not be fooled into thinking that just because a supplier does well on a survey quality hardware will result.

Maintaining Objectivity

When evaluating supplier quality issues, we strongly recommend maintaining a sense of objectivity. When the production line stops and the cause appears to be nonconforming supplier hardware, it's tough to be objective. As Deming teaches, look for solutions rather than someone to blame. Philip Crosby teaches that half of all supplier problems are ultimately found to be customer-induced. Sometimes the problem is an obvious customer inspection error (a customer finds a nonconformance, the supplier is called to the customer's facility, and the supplier shows the customer that the inspection was performed improperly).

We've all been embarrassed by these situations. Sometimes incorrect requirements are provided to the supplier, the supplier meets the erroneous requirements, and the supplier component doesn't work when installed in the customer's product. This problem relates to the supplier communication issues discussed earlier, and the importance of clearly defining requirements in purchase orders and on engineering drawings. Recall the situation at Bernard Medical Devices described at the beginning of this chapter. It's likely Bernard will find the requirements levied on its suppliers do not assure supplier materials will work in Bernard's blood chemistry analyzers. That's a failure at Bernard, not at its suppliers.

On the other hand, customer procurement specialists have to also be objective in their evaluation of supplier performance. Procurement specialists often feel responsible when suppliers fail to deliver quality hardware. We've found that customer quality assurance, manufacturing, and engineering personnel are often too quick to blame suppliers (as mentioned above). We've also occasionally found the opposite condition in customer procurement personnel, who are sometimes too quick to defend recalcitrant suppliers (some have even earned a reputation as "vendor defenders"). As stated in Chapter 5, the first step in solving a problem is to objectively define it, and all personnel should strive to do this without searching for someone to hang.

Summary

Most manufacturing companies spend more than half of their sales dollars on purchased parts and services. It's not at all uncommon for this percentage to be more like 70 or 80 percent of a company's total sales dollars. Even if an organization's quality management process could make everything done in house perfect, more than half the work would still be ignored. The missing key ingredient, one that many of us fail to consider, is the supplier base.

Suppliers have to be included in the quality improvement process. Successfully extending the quality improvement process to suppliers requires effectively communicating, making suppliers part of the development and continuous improvement processes, focusing on supplier cost structures (rather than supplier prices), emphasizing long term relationships with fewer suppliers, and selecting suppliers based on the results of objective evaluations.

References

Let's Talk Quality, Philip B. Crosby, Penguin Books, 1990.

Quality Without Tears, Philip B. Crosby, Penguin Books, 1984.

Quality Is Free, Philip B. Crosby, McGraw-Hill Publishing Company, 1979.

Quest for Quality, Roger L. Hale, Douglas R. Hoelscher, and Ronald E. Kowal, Tennant Company, 1989.

Well Made in America, Peter C. Reid, McGraw-Hill Publishing Company, 1990.

Competitive Advantage, Michael E. Porter, The Free Press, 1985.

Competitive Strategy, Michael E. Porter, The Free Press, 1980.

Thriving on Chaos, Tom Peters, Alfred A. Knopf, Inc., 1987.

The Improvement Process, H. James Harrington, McGraw-Hill Publishing Company, 1987.

"It Takes Hard Facts to Make Good Partners - How Industry Selects Suppliers," *Purchasing Magazine*, 1992.

Chapter 16

D1-9000, ISO 9000, MIL-Q-9858, and MIL-STD-1520

The standards for quality management....

There are four major quality management standards we believe any manager in a high technology manufacturing environment should understand. The four key standards are:

- MIL-Q-9858: *Quality Systems Management*

- MIL-STD-1520: *Corrective Action and Disposition System for Nonconforming Material*

- ISO 9000: *Quality Systems*

- D1-9000 (Revision A): *Advanced Quality System*

The first two of the above standards, MIL-STD-1520 and MIL-Q-9858, were issued in their original form several decades ago. There age notwithstanding, both are excellent quality management guides. ISO 9000 and Boeing's D1-9000 (Revision A) came later, and add to the formal requirements for managing a quality manufacturing system. We will begin our discussion with MIL-Q-9858 and its requirements.

MIL-Q-9858

MIL-Q-9858, *Quality Program Requirements*, was the government's first standard focused specifically on managing quality in a high technology, systems-oriented development and production environment. Prior to that, and for manufacturers who build simpler products, the government can impose MIL-I-45208,[5] which governs inspection requirements. MIL-I-45208, as a standard focusing on inspection requirements, was primarily intended for use in companies that rely on detection as a quality philosophy. MIL-Q-9858 was the government's first standard that had as its underlying foundation a shift toward a prevention oriented quality management system. Significantly, MIL-Q-9858 was first released by the government in 1958.

MIL-Q-9858 requirements are summarized below:

- *Quality Program Management*. MIL-Q-9858 contains requirements related to quality management organization, and it states that personnel involved in managing the quality function shall have the authority and organizational freedom to effectively manage an organization's quality.[6] It further defines a requirement for initial quality planning, which states that all parts of a contract should be reviewed as early as possible to assure all special processes and other requirements are incorporated into the organization's plans for managing the contract. MIL-Q-9858 goes on to state that adequate and documented work

---

[5] Note that MIL-I-45208 is still in use, and is typically imposed on government contractors who build simple items.

[6] Contrary to the position frequently taken by many government quality assurance representatives, MIL-Q-9858 does not require that a company's quality assurance manager report to the president. This part of the standard is often incorrectly interpreted as such a requirement.

instructions shall be created to assure that all actions necessary to execute a contract are performed in a manner consistent with contract requirements. The standard next states that quality records shall be created and maintained to allow subsequent reviews to determine that the product was built in a manner consistent with contract requirements. Within this section, MIL-Q-9858 also requires that contractors shall maintain a corrective action system for identifying and correcting adverse quality trends. This includes identifying the costs related to quality for both prevention and detection oriented activities.

- *Facilities and Standards*. In the first part of this section, MIL-Q-9858 requires contractors to maintain control over the drawing release and change process, with a special requirement for assessing the engineering adequacy of drawings prior to their release. MIL-Q-9858 requires that measuring and test equipment be capable of determining that the product conforms to engineering requirements. Within this section, MIL-Q-9858 also imposes MIL-STD-45662, which is the military standard for maintaining a measuring equipment calibration standard (the essence of MIL-STD-45662 is that any equipment used for assessing the acceptability of deliverable product must be calibrated to standards traceable to the National Institute for Standards and Testing).

- *Control of Purchases*. Recognizing that most contractors subcontract a large portion of their work (as mentioned previously in this book), MIL-Q-9858 establishes requirements for assuring adequate design and quality requirements flowdown to suppliers, supplier performance measurement, and supplier quality management systems. In simpler terms, MIL-Q-9858 requires that a contractors' suppliers meet the quality standards imposed by the government on the contractor.

- *Manufacturing Control*. MIL-Q-9858 requires that incoming supplies and materials be subjected to receiving inspection to assure compliance with design and quality

requirements. MIL-Q-9858 similarly requires in-process inspection and testing to assure that work in process meets design and quality requirements. MIL-Q-9858 requires that final inspection and testing be performed to assure that the final product complies with its design and quality requirements. Within this section, there are other special requirements. The first of these special requirements is that handling and storage do not degrade product quality. The next is that the contractor control any nonconforming items such that they cannot be accidentally mixed in with conforming product. Another set of special requirements pertains to statistical quality control requirements (these requirements specify different sampling approaches, and refer the contractor to other military sampling plans, most notably MIL-STD-105 for lot sampling, and MIL-STD-414 for continuous production sampling, for sample size and accept/reject criteria determination). This section of MIL-Q-9858 concludes by stating that contractors shall positively identify all items' inspection status, such that it can be readily and conclusively determined if an item has been inspected and it conforms with inspection requirements.

- *Coordinate Government/Contractor Actions*. The last section of MIL-Q-9858 specifies actions that should be coordinated between the contractor and the government. These requirements are related to government personnel access to the manufacturer's facility for inspection purposes and control of government-furnished property.

We believe MIL-Q-9858 is a superior requirement that has served the nation well for more than 40 years. It defines a logical set of requirements for managing in a high technology defense environment.

Within the last 10 years, there has been a move to shift defense contract requirements to ISO 9000, and to make MIL-Q-9858 obsolete. The U.S. Defense Department has stated that it intends to obsolete MIL-Q-9858 and instead rely on ISO 9000. That has not really occurred. Many of the MIL-Q-9858 requirements, with slightly modified language, have found their way into ISO 9000. That this is the case makes sense, as the basic MIL-Q-9858 requirements make sense. Many DCMAO (Defense Contract Management Area Operations, the renamed DCAS organization), in contravention to the government's stated intent, still feel that MIL-Q-9858 is the governing document. Accordingly, many of the government's DCMAO quality assurance representatives still object when a contractor departs from MIL-Q-9858 requirements.[7] Finally, some U.S. government procuring agencies are still imposing MIL-Q-9858 in their contracts. The bottom line is that MIL-Q-9858 is an excellent standard for managing a high technology defense contractor development and manufacturing enterprise, and contrary to the government's stated intent, it will most likely be with us for years to come.

MIL-STD-1520

In an earlier chapter, we developed the concept of a quality measurement system, and we made the point that in order for such a system to be effective, all nonconformances should be documented. The U.S. Defense Department realized this decades ago, along with many other requirements for dispositioning nonconforming material, controlling nonconforming material, and implementing corrective actions to prevent future nonconformances. MIL-STD-1520C, *Corrective Action and Disposition System for Nonconforming Material*, is the current document governing these areas for manufacturers who build products for the U.S. military.

MIL-STD-1520C provides several definitions and requirements for the material review function, and as mentioned earlier, a manufacturing organization's senior management Corrective

Action Board. Several definitions are pertinent to this discussion:

- *Disposition*. Disposition refers to the actions an organization can take with nonconforming material. Dispositions include rework, repair, scrap, use as is, return to vendor, and no defect. Each term is explained below.

- *Rework*. Rework means reworking a product such that it conforms to all design requirements. In essence, it means bringing the nonconforming item to a conforming state such that the nonconformance no longer exists. Ordinarily, a manufacturer would not need permission from the government to implement a rework disposition.

- *Repair*. Repair means taking actions to minimize or eliminate the effect of the nonconforming condition, but doing so in a manner that still leaves the item in a nonconforming condition. An example might be a cracked shaft that is subsequently welded to fix the crack. The weld is not part of the design, but it eliminates the effect of the crack. Unless a manufacturer has MRB authority for repairs,[8] government permission is required to implement a repair disposition.

- *Standard Repair*. A standard repair is one that is expected to occur relatively frequently and for which the government grants permission to use for a specified period of time or a specified number of units. For example, the government may allow a manufacturer to use jumper cables on a circuit card to repair broken traces for the life a contract.

- *Scrap*. Scrap means that the manufacturer makes a decision to scrap the nonconforming item. Manufacturers typically would not need

---

[7] The continued reliance on MIL-Q-9858 is a curious situation. One can only guess at the reasons. DCMAO personnel are civil servants, and civil servants have in the past been observed to be slow to react to change. Perhaps it is because the ISO standards are foreign. The most likely reason is that most DCMAO civil servants know MIL-Q-9858 well, they agree with its very logical philosophy, and they find it a useful tool for overseeing a defense contractor's quality assurance program.

[8] MRB authority can be granted at the government's discretion to allow a manufacturer to implement repairs or use as is dispositions without asking the government for permission in each instance. If such blanket MRB authority is not granted by the government, then permission must be granted each time the manufacturer wishes to implement a repair or a use as is disposition.

government permission to scrap a nonconforming item, unless the item was government-owned property.

- *Use As Is*. Use as is means that the manufacturer's engineering organization has determined that the effects of the nonconformance are insignificant, and the nonconforming item can be used in the nonconforming condition. In a manner similar to repair dispositions, manufacturers must ordinarily request government permission to use nonconforming items as is.[9]

- *Return to Vendor*. Return to vendor dispositions can be used when supplier components do not conform to design requirements. This disposition simply returns the nonconforming item to the supplier who provided it.

- *No Defect*. A no defect disposition means that the inspector who rejected the item did so in error, and there is no defect.[10]

- *Corrective Action*. Corrective action refers to the actions a manufacturer takes to eliminate the root causes of a nonconforming condition. Corrective action differs from disposition in a significant way. Disposition simply refers to how the nonconforming item is processed (rework, repair, scrap, etc.); corrective action refers to the actions taken to prevent the nonconformance from occurring again.

- *Preliminary Review*. Preliminary Review is a review of the nonconforming item, typically by a manufacturer's quality engineering

organization (sometimes the manufacturer's manufacturing engineering and manufacturing supervision also participate in the Preliminary Review). The Preliminary Review function can implement no defect, scrap, return to vendor, and rework dispositions. The Preliminary Review function should also identify corrective actions to preclude nonconformance recurrence.

- *Material Review Board*. Any items not dispositioned by Preliminary Review are referred to the Material Review Board, or MRB. MRB members typically include a manufacturer's quality engineering, manufacturing engineering, manufacturing supervision, procurement, and engineering functions. If the manufacturer has been granted MRB authority by the government, the manufacturer's engineering representative must approve any repair or use as is dispositions. If the manufacturer has not been granted MRB authority by the government and wishes to submit a request for a repair or a use as is disposition on a specific nonconformance, the manufacturer's engineering organization must approve the request before it is submitted to the government, and provide the analytical justification for the repair or use as is disposition with the request.

- *Corrective Action Board*. The Corrective Action Board (CAB) is comprised of an organization's senior management team and its function is to review data from the manufacturer's quality measurement system, identify dominant nonconformances and adverse quality trends, and assign teams to quality improvement projects (as discussed in the chapter on employee involvement and empowerment) to improve quality in specified areas.

Figure 16-1 shows the typical Preliminary Review and MRB process manufacturers who sell to the U.S. Defense Department follow.

MIL-STD-1520C also delineates specific requirements for identifying and segregating nonconforming hardware. The standard requires

---

[9] It is generally not a good idea to request permission to use a nonconforming item in the nonconforming condition if the item can be reworked to conform to design requirements. If the item can be reworked, most government quality assurance representatives will not approve a use as is disposition.

[10] In our experience, it always makes sense to discuss the "no defect" disposition with the inspector who initially rejected the item for three reasons: To educate the inspector, to give the inspector an opportunity to correct the person making the no defect disposition if the no defect disposition is incorrect, and as a courtesy to allow the inspector to know the ultimate disposition.

that all nonconforming items be identified as such, and that nonconforming items be segregated from conforming product. The intent is to prevent nonconforming product from being mixed in with good product and accidentally being shipped as good product.

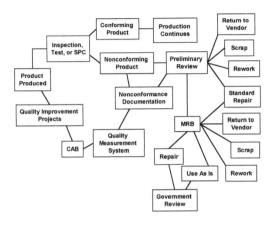

Figure 16-1. The Preliminary Review and MRB Process Outlined In MIL-STD-1520C. This process provides a great approach for dispositioning nonconforming hardware.

We will also point out that MIL-STD-1520 has also been declared obsolete by the U.S. government, but it is still routinely imposed by Defense Department procuring agencies and DCMAO.

ISO 9000

The International Organization for Standardization (ISO), a European body charged with creating and maintaining international standards, developed a set of standards for quality management. The overall body of standards is generically referred to as ISO 9000, but the standard is actually made up of several subparts as shown in Figure 16-2.

- *ISO 8402*. ISO 8402, *Quality - Vocabulary*, defines terminology used throughout the ISO 9000 series of specifications.

- *ISO 9004, Part 1*. ISO 9004, Part 1, *Quality management and quality system elements – Guidelines*, is somewhat similar to MIL-Q-9858 in that is outlines policy, organizational

responsibilities, and other related information and the requirements for an organization's Quality Manual.

- *ISO 9000*. ISO 9000, *Guidelines for selection and use of the standards*, does as its name implies. This portion of the standard provides guidance pertaining to which portions of the standard are applicable to an organization depending on the nature of the organization's business.

- *ISO 9001*. ISO 9001, *Quality systems - Model for quality assurance in design/development, production, installation, and servicing*, is the ISO standard that applies to organizations that both design and manufacture products. This standard is quite similar to MIL-Q-9858, in that it outlines requirements for contract review, drawing control, process documentation, and other requirements for assuring quality in an engineering development and manufacturing organization.

- *ISO 9002*. ISO 9002, *Quality systems - Model for quality assurance in production and installation*, is similar to ISO 9001 except that it does not include requirements for engineering development. ISO 9002 is intended for use by build-to-print manufacturers.

- *ISO 9003*. ISO 9003, *Quality systems – Model for quality assurance in final inspection and test*, focuses primarily on the testing and inspection of completed products. It would be applicable to organizations engaged primarily in test and inspection activities.

- *ISO 9004, Part 2*. ISO 9004, Part 2, *Quality management and quality system elements – Guidelines*, applies primarily to organizations that provide services rather than manufactured items.

U.S. interest in the ISO 9000 series increased in the last 10 years primarily because major aerospace prime contractors demanded its implementation, and also because many U.S. companies felt that foreign sales would increase if they were certified to the ISO 9000 standard. McDonnell Douglas, in the last few years before it merged with Boeing, required that all of its suppliers either become certified to ISO 9000 or have plans for doing so. After McDonnell Douglas merged with Boeing, Boeing's D1-9000 standard, to be discussed shortly, essentially incorporated the same requirement.

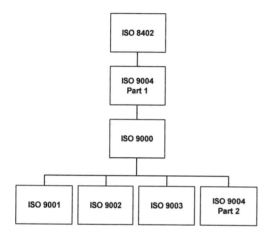

Figure 16-2. ISO 9000 Hierarchy. The ISO 9000 standard is made up of several supporting standards.

The ISO 9000 certification process is lengthy and expensive. It typically involves contracting with a consulting organization to guide the company's implementation process (which typically takes 12 to 18 months) and a series of pre-audits by the consulting company to assess the company's readiness. The company next contracts with another registrar consulting company for the actual initial ISO 9000 audit. If the company is deemed acceptable by the registrar, it is granted the ISO 9000 certification. The process is repeated periodically in follow-up audits, typically with the consulting company helping the company re-prepare for the follow-up audit, and another consulting company performing the audit. The consulting costs alone for the initial preparation and audit can exceed $100,000, depending on the company's size.

As you might infer from the above, the enthusiasm for ISO 9000 among U.S. companies has been a windfall for many consulting groups in the U.S. and Europe, and in fact, many consulting

organizations were created to meet the demand for ISO 9000 implementation and audit services.

Has the ISO 9000 enthusiasm provided its hoped-for returns? For some companies, the answer is yes; for others, it is no. Many companies have spent their time and money to become ISO 9000 certified and seen essentially no increase in overseas sales. Others have had overseas sales waiting until the company attained ISO certification, and they have realized significant sales increases once the certification was awarded.

With regard to improved product quality, it has been hard for us to see any improvement in the quality of any company's goods and services that can be linked to ISO 9000 certification. In the final analysis, the ISO 9000 standards define the administrative and documentation requirements necessary to manage engineering and manufacturing quality. The ISO 9000 standards do not, by themselves, provide any assurance of quality improvement or compliance with any measurable product quality attributes. That responsibility still lies with a company's management.

Boeing's D1-9000 (Revision A)

Suppliers who sell to Boeing (which is the dominant aerospace company in the United States, both for commercial aircraft and defense systems) will all ultimately have to meet the requirements of Boeing's Advance Quality System standard, D1-9000 Revision A. Boeing's quality management standard is undoubtedly the best in the business, and any manufacturer that complies with it has a sound quality management system.

Originally, D1-9000 imposed Boeing's concepts of quality management on their suppliers, and all suppliers were told they had to comply or be working to a plan to comply in order to continue doing business with Boeing. The quality management system covered all aspects of an organization's quality management system, with a heavy emphasis on statistical process control.

About three years ago, Boeing updated the D1-9000 to Revision A. The revision incorporated two major changes: The statistical process control requirements were included in a separate, second half of the standard, and the first half of the standard was rewritten to include ISO 9001. The first half of D1-9000 literally incorporated the ISO standard, plus additional requirements Boeing felt were necessary in an aerospace manufacturing environment. The original ISO requirements are included in the text, and Boeing's additional requirements are added, in italics, in the appropriate locations. The overall approach is extremely well done, and any manufacturer who complies with D1-9000 (Revision A) in effect complies with ISO 9000.

Boeing also actively assists its suppliers in implementing D1-9000 (Revision A). Boeing presents implementation seminars and provides on site support at no charge to its suppliers to assist in implementation. Boeing is committed to making the standard work, seeing improved quality from its suppliers, and holding its suppliers accountable for their quality. D1-9000 (Revision A) is not a hollow set of documents, and the implementation requirements being demanded by Boeing are showing positive returns both for Boeing (in improved supplier quality) and its suppliers (in lower scrap and rework costs, and more business with the world's largest aerospace company).

AS9000 – An Emerging Aerospace Standard

In recent months, consideration for adopting AS9000, *Aerospace Basic Quality System Standard*, as an aerospace industry quality standard has received increasing attention. AS9000 was developed by a group of U.S. aerospace prime contractors, including Allied-Signal, Allison Engine Company, Boeing, General Electric Engines, Lockheed Martin, McDonnell Douglas (prior to their merger with Boeing), Northrop Grumman, Pratt & Whitney, Rockwell Collins, Sikorsky Aircraft, and Sundstrand. Significantly, the U.S. government was not actively involved in the AS9000 standard's development. AS9000 was developed and issued under the auspices of the Society of Automotive Engineers.

The intent and concept behind AS9000 are similar to Boeing's D1-9000 (Revision A). The standard is based on ISO 9000, with 27 additional

requirements unique to the aerospace industry. The intent is to standardize and streamline many of the other aerospace quality management standards.

At this time, widespread aerospace industry implementation and imposition of the AS9000 standard has not occurred, although it is likely that it will at some time in the future. The standard is referenced here only for completeness, and to alert the reader that quality management standards in the high technology sector (particularly the aerospace industry) are continuing to evolve.

Summary

All of the above standards can greatly assist managers in high technology manufacturing environments. The MIL-Q-9858 and MIL-STD-1520 standards, while aged, provide keen guidance on how to manage quality and implement meaningful corrective action. Even those not working in the defense contracting environment would be well-advised to consider the requirements in both standards. In particular, the guidance provided by MIL-STD-1520 and its Corrective Action Board requirements can help an organization to appropriately focus its efforts and realize meaningful and measurable quality improvement. ISO 9000 can similarly help an organization, and its overseas marketing implications are potentially significant. Boeing's D1-9000 (Revision A) is the most meaningful and comprehensive of all standards, taking an organization considerably beyond ISO 9000. Any organization meeting D1-9000 (Revision A) requirements has a strong quality management system.

Quality management standards in the high technology sector are continuing to evolve, with the most recent developments occurring in the aerospace sector. Acceptance by government agencies and other major organizations in the aerospace and defense industries of these new standards (and procuring agency willingness to drop other long-established quality management standards) must be managed on a case-by-case basis to assure compliance with customer requirements.

References

The ISO 9000 Handbook, Peach, Robert W., CEEM Information Services, 1994.

ISO 9000, Rothery, Brian, Gower, 1993.

MIL-Q-9858A, "Quality Program Requirements," Department of Defense, 1963.

MIL-STD-1520C, "Corrective Action and Disposition System for Nonconforming Material," Department of Defense, 1986.

ISO 9001, "Quality systems – Model for quality assurance in design/development, production, and servicing," International Organization for Standardization, 1987.

Advanced Quality System, D1-9000, Boeing, 1996.

Chapter 17

On-Time Delivery Performance Improvement

6 P's for improved delivery performance...

Attaining acceptable delivery performance is the most significant manufacturing challenge faced by many organizations. Manufacturing Resource Planning (MRP) systems, when implemented without considering other delivery performance factors, often do not provide hoped-for delivery performance improvements.

Our organization has found that delivery performance shortfalls are most frequently driven by problems that fall into six areas:

- Production Capacity.
- Production Control.
- Productivity.
- Procurement.
- Process Robustness.
- Product Delivery Responsibilities.

We've found that companies suffering from poor delivery performance usually have problems in all six of the above areas. To understand how to find and fix the problems in each area, we need to first consider how MRP works.

Manufacturing Resource Planning

MRP is a computer simulation of the factory and its manufacturing processes. The MRP system is a comprehensive and interactive data base that includes information on each product's bill of material, the manufacturing process, inventory levels, purchased parts' lead times, setup and run times for each operation, and other information relevant to the manufacturing process.

In most manufacturing organizations, MRP runs each week to process new orders accepted or forecasted by the sales department and entered into the MRP data base. When MRP runs, it assesses how many of all required items are in inventory and in various stages of the manufacturing build cycle. Based on this assessment and the MRP system's knowledge of future demand, the system determines how many new orders should be initiated, how many purchase requisitions should be generated, and all other actions required to meet the demand for additional product.

MRP issues daily dispatch reports to the manufacturing, purchasing, and stockroom areas. These reports define the jobs that are present in each area and when each should be completed or issued. The dispatch reports are how the system communicates to the factory. They define which jobs should be completed (and when they should be completed) in order to ship manufactured goods on schedule. As each required action defined by the MRP dispatch reports is completed, personnel completing the action make electronic entries into the MRP system. These entries inform the system of the status of all orders.

MRP assesses, on a daily basis, the status of all material receipts and issuances from stock, the status of all manufacturing operations, the status of all purchasing activities, and the status of components that have been rejected. MRP knows this because, as outlined above, company personnel make electronic entries to status the build sequence.

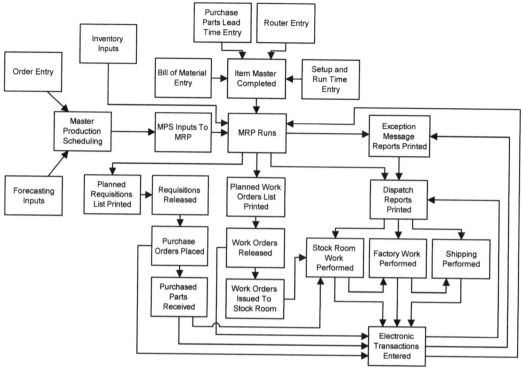

Figure 17-1. A Typical MRP System. Orders are entered, the system assesses build and inventory current status, and detailed work instructions are issued to the factory, the procurement function, and the production schedulers. If all dispatch report due dates are met, the product ships on time. If they are not, the system can deteriorate rapidly.

The MRP system also identifies detailed activities that are delinquent. As MRP receives the status information outlined in the preceding paragraphs, it compares the status of all actions to their planned status. MRP issues an exception message report to identify all actions that have not occurred on time. This report defines delinquent actions. Ideally, the exception message report should be used by the organization's schedulers and others to define areas of special focus. Figure 17-1 shows how a typical MRP system works.

In theory, the above process is elegant, and probably represents the best way to keep track of everything happening in a factory, especially if the factory has multiple work centers and multiple products. In practice, many organizations cannot keep up with the demands stated by the MRP daily dispatch reports (for reasons to be outlined below), and when they fall behind, delivery delinquencies occur. This paper focuses on what we have found

to be the more common delivery delinquency causes and how to correct them.

Production Capacity

We cannot overemphasize understanding and managing the lead time/capacity/load relationship. In our experience, organizations struggling with delivery performance often find this to be their most significant problem.

Over the past several years, manufacturing organizations and their leaders have come to understand that shorter lead times should result in lower manufacturing costs and improved customer satisfaction. Unfortunately, comprehension frequently stops at this point. Managers know shorter lead times are intrinsically better and they therefore want shorter lead times. Manufacturing organizations have to do things, though, to achieve shorter lead times. Simply quoting shorter lead times to customers without taking the necessary steps will not make customers happier or reduce costs (nor will it reduce lead times). Taking this route (promising shorter lead times without the capability to meet them) results in extremely dissatisfied customers, raises costs considerably, and induces delivery delinquencies.

MRP is the model that runs the factory. It is based on purchased parts lead times, manufacturing lead times, the manufacturing process, and inventory levels. The MRP system knows, based on these parameters, how long it takes to deliver product from order acceptance to shipment, and that should be the lead time the manufacturer quotes to its customers. Manufacturers frequently quote shorter lead times, though, because they want to please customers. Consider the following scenario:

"I need product in 16 weeks."

"I'm sorry, but our lead time is 22 weeks."

"But I really need it in 16 weeks."

"The best we can do is 16 weeks."

"If I can't get it in 16 weeks, I'll have to go elsewhere."

"Okay, we'll deliver in 16 weeks."

Does the above sound familiar? The problem with the scenario outlined above is that if the order is entered for delivery in 16 weeks and nothing is changed in the MRP model (and the factory/supplier network it represents), in 16 weeks the manufacturer will have accomplished little more than disappointing the customer to whom it made the 16-week delivery promise. Things will get even worse. The customer to whom the promise was made will not be the only disappointed customer. So will all the other customers whose orders go delinquent because of the 16-week commitment our manufacturer made to just one customer. How can that happen? If we only committed to an under-lead-time delivery to one manufacturer, should we not only go late to just that one customer, and not affect the others whose deliveries we booked at the correct lead

time? The answer is a resounding no. The likelihood is that if a manufacturer books just one order under lead time, the company is probably going to induce delinquencies for several other customers. To understand why, we need to turn to our next topic: Capacity.

Capacity is a measure of how much a factory and its suppliers can produce in a specified period. To understand it, we must think about the manufacturing process and the constraints associated with each step in the process. These constraints can best be defined through the use of capacity assessments that show how many standard hours of work can pass through the work center in a given period. To identify capacity, a manufacturer needs to know how many machines and people are available to perform work.

Simply identifying capacity is only half of the problem, though. Manufacturers also have to consider the load going through each of the work centers in the manufacturing process. This is where the MRP system helps. As outlined earlier, the MRP system defines how many jobs have to move through each work center. If the amount of time required for each job is identified through the use of standards (a standard is an estimate of the period of time it should take to perform an operation), then a capacity analysis can be performed to compare the amount of work each work center can perform to the amount of work it will have to perform in order to meet the dispatch report requirements.

If a manufacturer knows the standards for performing the operations that have to move through each work center and the capacity of each work center, the manufacturer can compare each work center's capacity to its load. If the capacity is greater than the load, the manufacturer should not have a problem (the work center can support the load, and the work should be accomplished on time). If the load exceeds the capacity, then the manufacturer has a constraint. The work center has more work than it can perform, and it will not complete the jobs it is supposed to in time to meet the dates specified in the MRP-generated dispatch report. At least a few of the jobs moving through the work center will fall behind. Unless the downstream work centers have excess capacity, it is not likely these jobs will recover to their downstream dispatch report due dates, and that means the delivery to the customer will be late.

We have found a number of ways in which the capacity versus load challenge can be inadequately considered by manufacturing organizations:

- The organization has inaccurate or no standards. If such is the case, capacity versus load considerations cannot be performed with any accuracy.

- The organization does not perform capacity analysis or performs capacity analysis infrequently.

- The organization has accurate standards and performs periodic capacity versus load assessments, but the findings are not considered in production planning.

Any of the above can be deadly to delivery performance, and we suspect that several readers will recognize that their companies suffer from one or more of the above problems. One might ask the question: How can a company operate at all if it does not accurately address capacity versus load? Many companies do, and they do so because they have excess capacity in other areas of the operation. This will allow such companies to fall behind in one area of the operation, but make up the lost time in subsequent downstream operations with excess capacity. Such situations frequently exist in poorly-managed companies, and during and after economic downturns (at least for a while after the downturn).

Where excess capacity exists, it usually exists because the company has not tailored capacity to meet market conditions. In these situations, companies can be lulled into believing that capacity versus load assessments are not necessary to assure delivery performance. Such a belief is dangerous for two reasons:

- From a cost containment perspective, the company should be concerned about excess capacity (the company is paying for capacity it does not need).

- The company may make delivery commitments on future orders, especially during an economic upturn, and find out too late that it does not have the required capacity to deliver in accordance with its commitments.

Let's consider the lead time versus capacity issue. What we need to consider is that lead time is directly influenced by capacity and load, and in reality, lead times are not fixed. They vary as the capacity and the load change. We will approach the discussion by recognizing two situations: One in which the plant and its work centers are operating below capacity, and the other in which at least one of the plant's work centers is operating at or above capacity.

If the organization is operating below capacity (i.e., it has excess capacity), product lead times are determined strictly by how long it takes to setup and run each job and to move jobs from one work center to the next. The idea here is that as a job moves into a work center, there is a machine available to set up and run it immediately. Lead time, in this situation, becomes the simple sum of the times required for the purchased parts, the setup and run operations, and moving the product between operations (with appropriate consideration given to operations that occur in parallel and those that occur in series, as indicated in Figure 17-2).

If an organization is at or over capacity, though, determining lead time becomes more complicated. There are two concepts we must consider:

- When the organization is at or over capacity, load and lead time are directly proportional. As the load in a work center increases, the lead time for each incoming job increases. Instead of being able to set up and run the job immediately, it must wait as other jobs are set up and run. The higher the load, the longer the lead time.

- When the organization is at or over capacity, capacity and lead time are inversely proportional. As the capacity in a work center increases, the lead time for work to move through that work center decreases. The higher the capacity, the shorter the lead time.

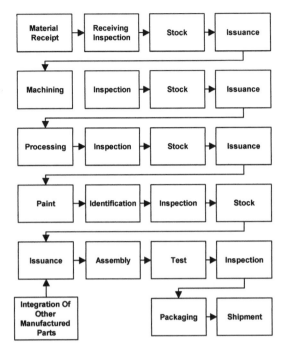

Figure 17-2. A Typical Manufacturing Process. The process can be thought of as a chain, with overloaded work centers being the weak links that constrain and therefore define the factory's output.

Most organizations operating at or above capacity address the above relationships (and their inherent limitations) by adding capacity internally or by offloading work, or through the use of buffers or queue times for each work center. Offloading work or adding capacity internally are self-explanatory solutions to this dilemma. Buffers or queue times are a bit trickier. They are a measure of the predicted amount of time a job will have to wait in a work center before it can begin the setup and run process. In effect, queue times are buffers that represent an implied understanding that jobs will not get on machines immediately. Queue times are the organization's estimate of how long jobs will have to wait. In this situation, capacity is reached when the load in the work center exceeds the sum of the setup, run, and queue times.

Readers who have worked in manufacturing will probably feel that most of the above is obvious, and to an extent, intuitive. Readers who have never worked in manufacturing (as is often the case for people in sales or marketing positions who make delivery commitments for the company), the relationship between lead time and capacity is neither intuitive nor obvious. We believe that manufacturing managers making delivery commitments have to understand the lead time/capacity/load issue. We further believe that manufacturing managers have to recognize that others outside manufacturing will not intuitively understand the lead time/capacity/load relationship, and that the desire to quote reduced lead times has to be carefully managed.

Given the above considerations, what are the things responsible managers can do to intelligently commit to delivery schedules? Here are our recommendations:

- Understand the organization's existing lead times, publish them, and do not allow the sales department to commit to earlier deliveries without the manufacturing organization's concurrence.

- Regularly assess capacity versus load in all work centers. We recommend performing this analysis on a weekly basis. Where loads exceed capacity, increase capacity (internally or through off loads) or lengthen the quoted delivery time.

- In cases where orders are booked under lead time (for marketing or other considerations), replan and micro-manage the progress of the work through the factory to assure the end item delivery is met.

Production Control

Production Control is the discipline that determines what needs to be built and when it needs to be built in order to ship product on time. Doing so requires real planning skills (the ability to work backwards from a future point in time to determine what has to happen and when it has to happen). Our observation is that with the advent of MRP systems over the last decade and a half, planning skills have deteriorated in many companies. This is perhaps a logical fallout of large scale MRP systems implementation.

Let us think about why this might be. In our earlier discussion in this chapter, we reviewed how MRP systems work and what they do. We explained that the MRP system, with its data base of component assembly times, manufacturing routers, and other information, identified when items had to be built and in what quantities to support required delivery times. Unfortunately, in many organizations, this MRP capability has resulted in production control and planning personnel who are, in essence, data entry clerks. The ability to truly plan, which has always been a rare attribute, has been made more rare by over-reliance on MRP systems.

The above situation might be acceptable if MRP systems had the capability of a human mind, but they do not. When work is not completed in a work center when scheduled, the MRP simply shows that it is past due and must be completed now. If more than one work order is late, it shows these work orders as late and directs that all delinquent work orders must be completed now. In short, MRP cannot account and plan around bottlenecks, rejections, items in MRB, or any of the other real-world situations manufacturers must contend with on an hourly basis. Doing so takes someone with the ability to plan.

To address this MRP shortfall, a company needs production control professionals who recognize when the data provided by MRP is no longer valid, and who can develop work-around plans to bring the company back on schedule.

Productivity

A manufacturing organization can have the best production control people in the world and adequate capacity, but if the plant's productivity is poor, work will not ship on time and ultimately, the plant will not be competitive.

There are numerous productivity measures. The productivity measure we have found best for integration with capacity and lead time issues is a measure of the actual time versus the standard time to perform a task. This is frequently referred to as efficiency, and is defined as:

Efficiency = actual time/standard time

To use this productivity measure, a plant has to have work standards for all tasks (or at least for most of the tasks performed during the manufacturing process). There simply is no way to get around this.[11] Some might view developing and having standards as a burden, but in our experience, without such standards a manufacturing organization is simply guessing at its costs, schedules, and lead times.

Why is the above so important? In addition to what should be a normal management concern (i.e., to assure all employees are working efficiently), we need to recognize that most MRP systems inherently assume that operations are occurring at 100 percent efficiency. If the plant is averaging less than 100 percent efficiency, it will run behind the MRP schedule, and delinquencies will result. Delivering manufactured goods on time will not occur without measuring efficiency and taking the necessary steps to assure that any inefficient areas are brought up to standard.

Procurement

"Acme Manufacturing, one of our suppliers, delivered late, and that's why our product is shipping late."

Have you ever heard the above? One of our more frequently-encountered explanations for delinquent deliveries is late purchased parts delivery. Most manufacturing organizations buy as much as 75 percent or more of their products from suppliers, so the potential for supplier failures certainly exists. Our experience indicates, however, that supplier failures frequently are not the reason materials are missing when needed. Usually, the failures are induced by the buying organization. We need to turn to a focused assessment of a typical procurement process (as Figure 17-3 shows) to understand this phenomenon better.

From the material planning and procurement organizations' perspectives, when MRP runs it checks the due dates of orders that have been

---

[11] For additional information on this subject and a PowerPoint Manufacturing Performance template, please contact our website at www.bhusa.com.

entered, the requirements based on the bill of materials, and inventory status. Based on these assessments, the system defines additional materials to be purchased and when they should be ordered based on supplier lead times (as previously input to the system). MRP provides a dispatch report of recommended requisitions. The planners or buyers should review this list and release the requisitions recommended by the system and their knowledge of ordering practices, likely future orders, and other factors. Once the requisitions are released, the buyers should place the orders with the delivery times the MRP system indicates it needs.

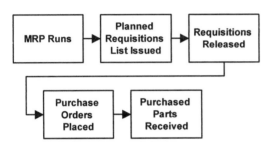

Figure 17-3. A Typical MRP-Driven Procurement Process. Logical performance metric points include requisition release, purchase order placement, and purchased materials receipt.

The above material planning and purchasing steps have "run" times just as manufacturing operations do. Many times, managers fail to monitor material planner and buyer performance in meeting these run times. Most MRP systems have inputs that tell the system how long it should allow for buyers and planners to review and release requisitions, convert them to purchase orders, and place the purchase orders.

We have frequently found that these internal procurement action "setup and run" times (i.e., the time to convert the requisition to a placed purchase order) are violated. If the buyers and planners take too long to accomplish these actions, they may do the same thing the Marketing people frequently want to do, and that is to violate lead times. When this occurs, the supplier lead time is violated, and the supplier is likely to deliver late.

Another problem we frequently encounter is purchase orders with due dates that do not support the MRP need date. This means the supplier may deliver on time (i.e., meet the purchase order due date), but the material will still not arrive on time to allow an on-schedule end-item delivery. The reasons for this can include internal excesses as outlined above (taking too long to place the purchase order) or changing market conditions that increase supplier lead times (as is currently occurring for titanium, forgings, extrusions, etc., due to the aerospace industry upturn). Organizations with poor delivery performance that track supplier delivery performance and show a high percentage of on-time supplier deliveries often have this problem. The suppliers are on time, but their deliveries do not support the procuring organization's MRP need dates.

In yet other instances, purchase orders are simply not placed. Buyers and material planners make mistakes. Our recommendations to address the procurement issues outlined above include:

- Define and publish internal lead times for planned requisition review, requisition release, and purchase order placement.

- Develop a report that shows all instances in which the above lead times are being violated, and identify and correct the causes of the violations. We recommend developing these reports and tracking the data from both company and individual buyer and planner perspectives.

- Develop a report that shows all instances in which purchase orders have due dates that do not support the MRP need date, and identify and correct the issues inducing such non-supporting purchase orders. We recommend developing this report and tracking the data from both company and individual buyer perspectives.

- Develop a report that shows all unplaced purchase orders. We recommend developing this report and tracking the data from both company and individual buyer perspectives.

- Constantly track supplier lead times and immediately modify the MRP data base to show changes as they occur.

While the above actions may seem intuitive, we are often surprised at how many nonsupporting supplier deliveries are induced by procurement (and not supplier) failures. We strongly recommend taking a hard look at the procurement function; it is an area of low-hanging fruit for improving delivery performance.

Process Robustness

As mentioned above, MRP systems generally assume all processes are robust; i.e., rejections will not occur. MRP systems do not make allowances for rejections in their planning. When a component is rejected, the time it takes to rework, repair, or replace the component is not included in the MRP routers and their associated lead times. That means that each rejection carries with it a requirement for significantly increased work-order-specific planning, and a much higher risk the item will ship late.

Preventing nonrobust processes from inducing delinquencies requires an aggressive failure analysis and corrective action approach, as well as superior planning to develop rapid recovery plans. We recommend the problem-solving and systems failure analysis approach outlined earlier in this book. The problem-solving and systems failure analysis approach outlined in preceding chapters has worked well for manufacturers that have adopted it. The approach supports D1-9000 (Revision A), ISO 9000, MIL-Q-9858, MIL-STD-1520, and other quality management requirements.

Perhaps the most important considerations regarding process robustness are that processes should have high yields, but when rejections occur, they should be worked aggressively to prevent the rejection from influencing required delivery dates.

Product Delivery Responsibilities

While MRP has significant capabilities and it has helped organizations improve, the MRP concept has a few disadvantages. One disadvantage is that the MRP system's data-processing-nature and dispatch-report "to-do-list" outputs tend to drive companies to organize along manufacturing-process-based lines.

For single-product companies, this may not be a problem. As product variety increases, however, process-based, work-center dispatch reports tend to grow in terms of quantity of work orders and types of part numbers. The result is that how the product comes together becomes opaque to the human beings in the system. There are so many parts in the system (and so many different assemblies they go into) that no single person can sense how the schedule and the parts integrate to allow finished assemblies to ship on time.

The MRP system knows how the product is supposed to come together, but it does not know what has to be done when anomalies occur. That takes a human being. When this problem is compounded with the other problem we mentioned earlier (production control personnel who are data entry clerks), what happens is not good. Products don't ship on time.

The result of the above is that the MRP system, in a very real sense and in many companies, takes over all or nearly all of the thinking required to ship a product on schedule. If the organization and all of its suppliers are on schedule to their dispatch reports, and if the organization has not exceeded its capacity in any areas, a company can live with this situation. The word "if" as used here is a powerful qualifier, though. Organizations and their suppliers are usually not 100 percent compliant to the dispatch reports, and that is when another problem emerges: Unfocused delivery responsibilities.

Consider this common situation: Supplier deliveries are late, Marketing is making commitments to ship products below standard lead times, several of the work centers are not keeping up with their dispatch reports, some of the parts are rejected during manufacture, and one or more of the work centers is overloaded.

In a company organized along process (rather than product) lines, other than the Vice President of Manufacturing, who is responsible for shipping product on time? Which of the jobs in the work centers that are past due should be worked first? Which parts of the many that are in work are needed to finish a product so that it can be shipped on time? Which of the delinquent supplier parts need urgent attention? Who is working the rejected parts, and in what order, so that they can be reworked or remade and the products that need them can be shipped? With all of the above occurring, who can predict when the products will ship?

The above situation defines the essence of what occurs in many manufacturing operations that run on MRP systems. The questions are:

- Who untangles the situation?

- Who is the product champion who sees to it that products ship on time?

Our experience indicates that when companies are organized along process (and not product) lines, delivery performance failures are likely because the problem becomes too complex for a single person to solve.

Our recommendation is to organize the factory along product (rather than process) lines to the maximum extent possible. We recommend having all of the work centers unique to specific product lines report to individual operations managers responsible for the product lines. We recommend having other factory areas that provide generalized support (such as a machine shop, process lines, stock rooms, and other generalized functions) report to a single factory manager. We further recommend having the product-unique functions for each product area (e.g., manufacturing engineering, scheduling, final assembly, and any other product-unique manufacturing areas) report to the product-line-specific operations managers. Figure 17-4 shows a recommended organizational approach for a company with four product lines.

The above approach provides for single individuals to champion product delivery performance, to act as a magnet to draw required parts into and through the plant, and to resolve issues related to their assigned products.

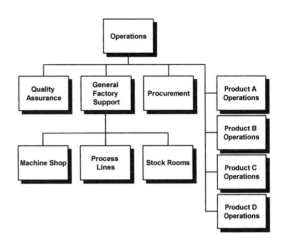

Figure 17-4. Recommended Operations Organization. This approach incorporates product-line-specific operations managers to champion on-time delivery for their product lines.

Summary

Our experience indicates that in MRP-based manufacturing organizations, delivery delinquencies are systemically driven by failures to understand and abide by lead times, failures to address capacity constraints, ignoring manufacturing productivity, diffused organizational responsibilities for on-time delivery performance, internal procurement failures, and nonrobust processes. Our recommendations for delivery performance improvement include understanding the nature of the capacity/load/lead time relationship, developing meaningful lead times, and only departing from lead times with supporting reschedules and focused management.

We recommend that manufacturers optimize the robustness of their processes using a systems failure analysis process focused on rapid cause identification and corrective action implementation.

We believe manufacturers have to understand their productivity and act on areas not meeting standard.

Manufacturers should understand the procurement process and its associated *internal* lead times, and monitor the procurement organization's performance to assure supplier commitments and deliveries that support MRP need dates.

Finally, we recommend organizing operations along product (rather than process) lines to assure appropriate focus on delivery performance.

References

Delivery Performance Improvement, Berk, Joseph H. and Berk, Susan H., Managing Effectively Seminars, Upland, California, 1997.

Managing Effectively, Berk, Joseph H. and Berk, Susan H., Sterling Publishing Company, New York, 1997.

"Systems Failure Analysis," Berk, Joseph H., *Proceedings of the Technical Program - Nepcon West '95*, Anaheim, California, 1995.

The Goal, Goldratt, Eliyahu M. and Cox, Jeff, Penguin, St. Paul, Minnesota, 1986.

Manufacturing Planning and Control Systems, Vollman, Thomas E., Berry, William L., and Whybark, D. Clay, Irwin, Boston, Massachusetts, 1992.

CASE STUDY 1

The SLAP Designation Pointing Error

Background

General Digital Electronic Systems Company developed and manufactures the Suspended Laser Acquisition Pod (SLAP). The SLAP is a helicopter-borne surveillance, target acquisition, and target designation system (see Exhibit 1). The system consists of a sensor suite that includes a thermal imaging sensor, a television sensor, a laser rangefinder and target designator, and associated environmental control, vibration-isolation, and other subsystems to support the sensors. The system is mounted in a cylindrical housing situated below the host helicopter's fuselage. General Digital is currently building its eighth production lot of Suspended Laser Acquisition Pods. Hundreds have been delivered to the U.S. Army.

Suspended Laser
Acquisition Pod

Exhibit 1. Helicopter-Borne Suspended Laser Acquisition Pod. The SLAP system is housed in the cylindrical case beneath the helicopter.

Electro-optical target acquisition and designation systems such as SLAP have proven to be an invaluable asset to modern tactical air and ground forces. These systems provide passive target acquisition capabilities, which allow for platform survivability against anti-radiation missiles and other countermeasures. The laser rangefinding and target designation features of modern electro-optical systems provide extremely accurate ranging information and target designation capabilities to support a variety of smart munitions. The Apache helicopter TADS, the OH-58 SLAP, and the F-16 LANTIRN (all of the preceding are systems with similar capabilities) greatly enhance the mission effectiveness of U.S. military forces.

SLAP Operation

In operational use, helicopter pilots use the SLAP television sensor to detect targets during daylight conditions. The thermal imaging sensor is similarly used to detect targets during night operations (the thermal imaging sensor relies on the targets' heat signatures to produce an image for display to the pilot). Once a target is detected, the pilot can energize the SLAP laser to determine the range to the target, or to "paint" a laser spot on the target to provide a homing point for laser-guided munitions.

The Persian Gulf War

When Iraq invaded Kuwait in August of 1990, the United States deployed a significant military force to the Persian Gulf. Key elements of the Desert Shield deployment included the SLAP-equipped fleet of OH-58 helicopters. Prior to deploying overseas, the U.S. Army evaluated the combat readiness of its complex weapons systems. One such evaluation involved the Suspended Laser Acquisition Pod.

Combat Readiness Findings

During the pre-Persian-Gulf deployment SLAP system evaluation at Yuma Proving Grounds in southern Arizona, a disturbing finding emerged. The SLAP line of sight did not track with the laser line of sight, which resulted in a designation pointing error. The difference was significant (the SLAP designation pointing error exceeded specification limits). The nonconformance had a very real combat meaning: The deviation from requirements was enough to induce a miss for laser guided munitions. The result would, at best, mean degraded combat readiness. At worst, it could allow the target to defeat the launch platform. Clearly, a systems failure analysis was required to identify the root cause of the designation pointing accuracy degradation, and to recommend appropriate corrective action.

SLAP Failure Analysis

General Digital quickly assembled a team of top technical experts to identify the cause of the designation pointing error anomaly. The team initiated a fault tree analysis within the first month of the investigation (the top sheet of the fault tree is included as Exhibit 2 to this case study).

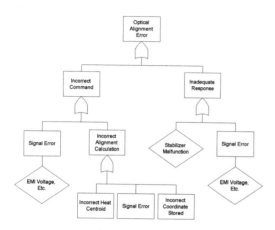

Exhibit 2. SLAP Fault Tree Analysis. The team prepared this fault tree during the SLAP alignment error failure analysis.

The failure analysis team initially focused on the SLAP boresighting system, as the team intuitively felt that a designation pointing error would most probably be related to a deficiency in the boresighting system. (Note: The boresighting system is used to align the SLAP laser, thermal imaging sensor, and television sensor to each other, and to the system line of sight.) After concentrating on the boresighting system for an extended period without finding the root cause of the designation pointing error (although other system problems were uncovered and corrected), the failure analysis team focused their efforts elsewhere. Detailed investigations into issues related to the SLAP power supply and the system microprocessor were pursued; however, neither yielded the root cause of the designation pointing error.

One of the problems faced by the failure analysis team included evaluating reams of technical and test data. Several months into the analysis, the failure analysis team recognized that the system test data showed that the laser rangefinder/designator

and the television sensors met their designation pointing error specification requirements. Only the thermal imaging sensor exceeded its designation pointing error requirements.

SLAP Optical Train

The SLAP sensors and target designation features communicate optically with the outside world through two windows, as shown in Exhibit 3. The first window is the thermal imaging sensor window, which is made of germanium. Germanium is an opaque black material (germanium is transparent to the thermal imaging sensor's infrared wavelength, however). The laser and television sensors see through the second window, which is transparent in the visible spectrum (the laser and the television sensors share many other elements of the SLAP optical train).

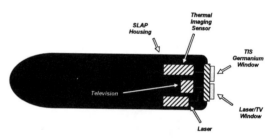

Exhibit 3. SLAP Internal Layout and Windows. Only the laser, television, and TIS are shown (other components are omitted for clarity).

Both the laser/television and thermal imaging sensor windows incorporate heaters to prevent fogging during system use. The thermal imaging sensor window has a heater that surrounds the window periphery. The laser and television sensor window has a very thin metal coating across the entire window surface. The metal coating is thin enough to be transparent, yet thick enough to conduct electrical energy and heat the window through resistance heating.

After additional analysis and confirming experiments, the failure analysis team concluded that the two windows' heaters had significantly different optical effects on their respective windows. The laser/television window heated the window evenly (due to the uniform metal coating used as a resistive heater). The thermal imaging sensor's germanium window heater did *not* heat its window evenly. Due to the thermal imaging sensor window's peripheral heating approach, the window experienced uneven heating (the outer edge of the window was heated more than the center). The effect was that during heating, the thermal imaging sensor window distorted, shifting the optical path away from that of the laser and the television.

The failure analysis team did not immediately understand why this problem had not surfaced earlier. Upon further investigation, the failure analysis team discovered that the window heater control software had been modified after completion of the SLAP development program. The software modification (directed by the customer) changed the temperature at which the window heaters triggered. Prior to the customer-directed change, the window heaters energized at an ambient temperature of 71°F (or lower), and maintained the windows at 71°F. After the software that controlled the window heaters was changed, the window heaters turned on at 105°F (or lower), and maintained the windows at this higher temperature. Note that the SLAP window heating feature was not evaluated during normal system acceptance testing at the General Digital manufacturing facility in Arcadia, California.

Failure Analysis Conclusions

After spending several months and significant funding to analyze the designation pointing error system failure, the failure analysis team identified the window heater control software as the root cause of the failure. Corrective actions included modifying the window heater software such that the window heaters held the window temperature at 71°F (as had been the case in the original configuration), which eliminated the designation pointing errors.

Questions

1. Systems failures occurring during production involve minor interruptions to the build cycle. Some involve temporary production interruptions. Occasionally, production lines are shutdown completely until the failure can be corrected. When such failures occur, the cost to a company can be significant. In this case, the SLAP designation pointing error failure was not detected until the system had been fielded and was ready to enter combat. What are your thoughts on the relative financial and combat readiness impact of the SLAP designation pointing accuracy failure?

2. What might General Digital have done differently to discover the system failure prior to hardware delivery?

3. What comments do you have on the four-step problem-solving process as it applies to the designation pointing error investigation?

4. What comments do you have on the top tier of the designation pointing error fault tree analysis? Should the top events be modified? Would such a modification have shortened the failure analysis? Why?

5. What are your comments on the failure analysis team's initial focus on the boresighting system?

6. What other analytical tool might have alerted the failure analysis team to the root cause of the designation pointing error anomaly?

CASE STUDY 2

Circuit Card Defects and Quality Measurement

Background

MMM Electronics develops and manufactures electronic systems for the U.S. government. The company specializes in electronic "boxes" that are used in larger systems, primarily on aircraft and ships. MMM Electronics has circuit design, software development, and manufacturing capabilities. The company employs approximately 150 people, with 24 in quality assurance, 18 in engineering, and 75 in manufacturing. MMM Electronics' remaining personnel are assigned to the program office, manufacturing engineering, procurement, planning, and other organizations.

MMM Electronics developed and currently manufactures the ON-149/C encryption/decryption device. The ON-149/C assembly is a shipboard system used to encrypt and decrypt communications signals between ships at sea and communications satellites. It is an electronics box consisting of a die-cast frame and cover assembly (procured from a die-casting/machining supplier) and seven circuit card assemblies (these include the processor board, the signal conditioning board, the power supply board, the annunciator board, the switching board, the output board, and the encrypt/decrypt board). MMM Electronics manufactures the seven circuit cards using pick-and-place, wave soldering, and touch-up operations. MMM Electronics also integrates the ON-149/C and performs final acceptance testing. The ON-149/C has been in production for approximately two years. Although the current production contract extends for only another nine months, the MMM Electronics management team expects that ON-149/C sales will continue for at least the next three years.

The Failure Analysis Assignment

Timothy R. Edwards is an engineer assigned to the ON-149/C program. Shortly after arriving at work one morning, his supervisor (Irving M. Wright), called Edwards to his office to discuss the ON-149/C program. Edwards recognized immediately that Wright was agitated as he began to describe the problem.

"We're having a significant number of ON-149/C components returned from the fleet," Wright said, "and we need to find out why."

"Do you have any information on the failures?" Edwards asked.

"Not really," Wright answered, "so I want you to assemble a team of the right people to get to the bottom of these failures. The only thing I know is that Sirahsira, the marketing guy, came back from Crystal City yesterday and he heard that the Navy is upset about the number of problems in the fleet. He's sweating bullets about the next contract. I want you to put a team together today and tell me what we need to do to fix the problem. See me at the end of the day and tell me where you are on this."

"Can I talk to Kurt to see if he has any additional insights into the Navy's concerns?" Edwards asked.

"Who?" Wright answered.

"The marketing guy," Edwards said.

"You could if he had a phone on the plane," Wright answered. "He's on travel again, this time to Europe. You probably won't be able to talk to him until tomorrow or the next day. See if his secretary can leave a message at his hotel in Paris."

The Failure Analysis Team

Edwards left Wright's office and immediately began thinking about how to approach the problem. He knew he should get Quality Assurance and Manufacturing involved. Edwards thought it would be a good idea to get someone from testing involved, too. He developed a preliminary list of names to include on the team Wright asked him to assemble. Edwards' list included:

• C. A. (Charlie) Rehquest (Quality Engineering)
• Samuel M. Blee (Manufacturing Engineering)
• Norman S. Trewmint (Testing)

Edwards returned to his office, called each of his prospective team members and invited them to a 9:00 a.m. meeting, and thought about what he might do to learn more about the problem prior to the meeting. He noticed the company's monthly quality summary report in a pile of documents on his desk, and began leafing through it to see if the quality data offered any insights into the ON-149/C failures.

MMM Electronics' Quality Measurement Data

MMM Electronics' quality management system conforms to the requirements of MIL-Q-9858 (Quality Program Requirements), and as such, the company tracks nonconformances for trends and corrective action adequacy. As Edwards studied the company's quality summary report, he observed that the MMM Electronics' Cost of Quality reports showed increasing costs for ON-149/C internal and external failures. The number of returned components from the fleet had increased in the last month, as had circuit card rejects during ON-149/C final acceptance testing. In particular, Edwards noticed two Pareto charts in the MMM Electronics quality summary report. (These charts are included as Exhibits 1 and 2. Exhibit 1 shows a Pareto chart for ON-149/C final acceptance testing failures. Exhibit 2 shows a similar Pareto chart for ON-149/C customer returns.)

After reviewing both Pareto charts, it became apparent to Edwards that the signal conditioning board had been the most-frequently-occurring nonconforming item (both within MMM Electronics and in the fleet). Edwards found additional information in another Pareto chart included in the MMM Electronics quality summary report.

The third Pareto chart provided data on the most-frequently-occurring signal conditioning board nonconformances within MMM Electronics. (This chart, included as Exhibit 3, shows the Pareto chart for pre-delivery signal conditioning board failures.) No data were available for signal conditioning boards returned from the fleet (other than those that

showed quantities of boards returned to MMM Electronics).

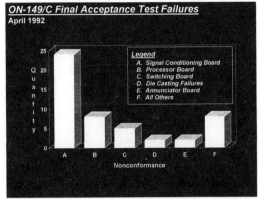

Exhibit 1. MMM Electronics' Internal (Pre-Delivery) ON-149/C Acceptance Testing Failures.

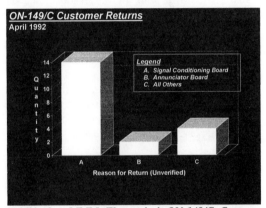

Exhibit 2. MMM Electronics' ON-149/C Customer Returns.

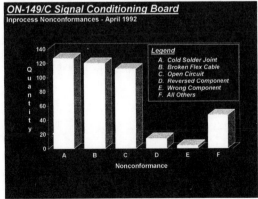

Exhibit 3. ON-149/C Signal Conditioning Board Nonconformances.

The Failure Analysis Team Meeting

At about 9:00 a.m., the three other failure analysis team members arrived in Edward's cubicle to discuss the ON-149/C failures. Edwards explained the problem as it had been relayed to him by Mr. Wright. Edwards also presented his initial conclusions on the signal conditioning board and its dominant failure causes.

"Based on the Pareto charts included in the quality data, it's apparent to me that the signal conditioning board is the dominant problem in house," Edwards explained to the group. "Does anyone have any different information, or does anyone know anything about the boards returned from the fleet?"

Rehquest, the quality engineer, answered first. "That seems to fit with my observations, and the Pareto chart on the signal conditioning board shows the situation about as I see it. Most of the production failures are either open circuits, cold solder joints, or broken flex cables. The flexes are a real pain to work with. I've seen a few solder bridges that caused short circuits, but they're noise level. Most of the open circuits usually end up being caused by either the cold solder joints opening up, or cracks in the flex cable."

"Yes, that's true," said Sam Blee, the manufacturing engineer. "Our operators are pretty careless when they work with these boards, and broken flexes are not unusual. Pride of workmanship just doesn't seem to be a factor down there."

"I agree with your conclusions about the broken flexes, Sam," Trewmint said. "Most of the system test failures go back to the signal conditioning board, and we see a lot of broken flexes. They're pretty fragile."

"What's the dominant failure cause on the boards during system testing?" Edwards asked Trewmint.

"I don't know for sure," Trewmint answered, "but it's probably the broken flex cables. We just write them up as a signal conditioning board failure and send them to the Material Review Board. We don't do any failure analysis in my area."

"Charlie, what happens after the boards go to MRB?" Edwards asked.

"Depends," Rehquest answered. "From MRB, we send them to the failure analysis lab. If the flex is broken, we scrap the board. If the flex is okay, the boards can usually be reworked."

Edwards spoke next. "Isn't that expensive, Charlie? I mean, scrapping the entire board if the flex is broken?"

"I guess," Rehquest answered. "But not all of them have broken flexes. Quite a few of the boards have solder joints that open up. We just send those back to solder touch-up."

"We see an awful lot of broken flexes in production," Blee added, "but not all of them are written up. Most of them are scrapped and we just put a new board through."

"You don't document those nonconformances?" Rehquest asked.

"Not if we're going to scrap them and issue a new board," Blee answered. "There's no rework required if we scrap the board, so why document the failure? Besides, we're not working on any of these right now, anyway. We've run out of flex assemblies, and the vendor can't deliver any more until the first of next month. For the time being, the issue of whether we need to write the broken flex cables up is academic."

"Let's get back to the boards that fail during system testing and don't have broken flex cables, Sam," Edwards said. "What happens to those boards next?"

"Well, after they're dispositioned in the Material Review Board, they go to the solder touch-up area for rework. After the boards are reworked, they're inspected by the touch-up area inspector," Blee answered, "and from there, they can go either back into production or to stock."

"Why would we put a board in stock instead of using it in production?" Rehquest asked Blee.

"For spares, I guess," Blee answered. "I believe this is one of the boards the Navy spares, because they do a lot of repairs during a cruise."

The group fell silent for a few seconds after Blee's last comment. Rehquest spoke next.

"How do we make these boards?" Rehquest asked Blee.

"It's a pretty straightforward process," Blee answered. "The board's a rigid-flex-rigid board, which means it actually consists of two circuit cards connected by a flex cable. Do you have a drawing here?"

"No, not at my desk," Edwards answered.

"Okay," Blee continued, "we can make do without one. The assembly people use a manual pick-and-place to install the components, and from there the board goes to wave soldering." Blee paused for an instant, lost in thought. "You know," Blee continued, "I'm really surprised we don't see more parts improperly installed or missing, considering the other workmanship problems we're seeing and the manual nature of the pick-and-place operation." Blee stopped for a second and then continued. "Well, anyway, after wave soldering, the boards are inspected, the defects are marked, and they go to solder touch-up. We have our best people in solder touch-up."

"How many of the boards have to go to solder touch-up after wave soldering?" Rehquest asked Blee.

"All of them," Blee answered. "Usually 10 to 20 percent of the solder joints require touch-up operations."

"Why?" Rehquest asked.

"Cosmetics, mostly," Blee answered. "The government will reject anything that looks like a cold solder joint, so we just catch it first and rework the questionable joints. Most of them are acceptable, but to keep the government happy, we do the cosmetics."

"You know, I really don't have much of a background in wave soldering," Edwards said. "Exactly what is it, and how does it work?" Rehquest and Trewmint leaned forward when Edwards asked the question. Edwards sensed that neither Rehquest nor Trewmint were familiar with the wave soldering process. All three listened carefully to Blee as he started to explain the wave soldering operation.

"It's very simple, really," Blee explained. "What happens," Blee began, "is the board, with all of the components inserted, goes on a conveyor. The component terminals, you know, the wires sticking through the board, are all facing down. The board gets carried by a conveyor across a solder bath, which flows over a little dam. The little dam makes a wave in the solder, see, that sticks up just high enough to let the solder hit the ends of the components. That's how they get soldered in place." With that, Blee went to the white board in Edwards' office and drew a picture to amplify his explanation. (The picture is included in this Case Study as Exhibit 4).

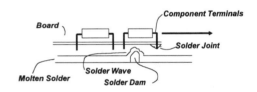

Exhibit 4. Sam Blee's Sketch of the Wave Soldering Process.

"That process seems like a delicate operation," Edwards said. "Couldn't it be the root cause of a lot of cold solder joints?"

"You bet," Blee answered. "We've got to control a lot of things, like how far the components are inserted into the boards, how much wire sticks out, how high the solder wave is, the solder temperature, the cleanliness of boards and the components, that sort of thing."

"Do the instructions the people on the floor work with spell this all out?" Trewmint asked. "You know, I used to be an hourly worker before I came to MMM, at my other company, and I always had to guess at what we were supposed to do. I worked in the circuit card production area before wave soldering was available, and our operation was pretty bad. We basically had to make our own tooling, and the assembly instructions said to make the cards like the engineering drawings. What kind of detail do our instructions provide here at MMM? And what about the tooling? If we're getting flex cable failures, doesn't that mean the cards aren't fixtured right during assembly and transportation between the work stations?"

The room fell quiet again, as Blee, Edwards, and Rehquest thought about Trewmint's question.

Questions

1. As a MIL-Q-9858-qualified defense contractor, MMM Electronics has a quality measurement program in place. Is MMM Electronics using its quality data effectively? What recommendations can you offer to improve the usefulness of this data at MMM Electronics?

2. What deficiencies exist in the quality data in the MMM Electronics quality reports? What do you recommend MMM Electronics do to correct these deficiencies?

3. Tim Edwards convened a team to review the ON-149/C failures and plan an analysis. Should anyone else have been included as a member on the ON-149/C failure analysis team? Can you offer specific recommendations for other team members?

4. What else could Tim Edwards have done to be better prepared for the initial failure analysis meeting? What else could he have done during the meeting to take advantage of the information that

emerged during the meeting participants' conversation?

5. What are your thoughts on the level of expertise and process knowledge displayed by the meeting participants?

6. Based on Rehquest's and Blee's comments in the failure analysis meeting, what conclusions can you draw about the adequacy of MMM Electronics' nonconformance documentation and quality data collection?

7. Based on your review of Exhibit 3, can you draw any additional conclusions about the adequacy of MMM Electronics' nonconformance documentation?

8. What are your thoughts on signal conditioning board workmanship at MMM Electronics?

9. What conclusions can you draw about the tooling, planning, and operator training at MMM Electronics?

10. Why do you think MMM Electronics is experiencing a signal conditioning board flex cable shortage?

11. Can an untested signal conditioning board go directly into the ON-149/C system? What are your recommendations related to this issue?

12. Based on the conversation occurring during the initial failure analysis meeting, how would you define the ON-149/C signal conditioning board problem Edwards' team is attempting to solve?

CASE STUDY 3

Laser Optics Debonding

Background

The Pave VIPER target designation and navigation system is used on a classified weapons platform. Pave VIPER is a very complex sensor and navigation system that provides the aircraft with night terrain following capabilities, as well as target acquisition (using television and thermal imaging sensors) and target designation (using a laser/transmitter) features. The system utilizes a dual-mode laser/transmitter capable of operating in the standard tactical wavelength (1.06 microns) as well as an eye-safe wavelength (1.54 microns, which is used for training). Portions of the Pave VIPER laser/transmitter optical path utilize independent optics (unique to each operating mode), while other portions of the optical train are common to both the tactical and training operating wavelengths. As a result of the added eye-safe feature, the Pave VIPER laser/transmitter has a higher-than-typical number of optical elements (most single-band tactical laser/transmitters have fewer than fifty optical elements; the Pave VIPER laser/transmitter has over 90 optical elements).

Omega Laser Systems in Maitland, Florida, developed the Pave VIPER laser/transmitter in the mid-to-late 1980s. The Pave VIPER laser/transmitter is a subsystem of the Pave VIPER system developed and manufactured by a major electronics systems integrator. The Pave VIPER system has been in production for about two years.

The Laser/Transmitter Production Process

Manufacturing the Pave VIPER laser/transmitter is a complex process consisting of numerous tests and frequent rework. Important elements of the laser/transmitter production process include:

- Mounting the optical elements (i.e., the lenses, prisms, and other laser energy focusing and steering optics) on titanium mounts.

- Adjusting these optical elements as they are installed in the laser/transmitter such that the laser beam is correctly focused and aligned.

- Environmentally testing the laser/transmitter to assure system integrity prior to delivery to the Pave VIPER systems integrator.

These three portions of the laser/transmitter production process are further explained below.

The Optical Element Bonding Process. Each of the Pave VIPER laser/transmitter optical elements is bonded to its titanium mount with a two part epoxy, as shown in Exhibit 1.

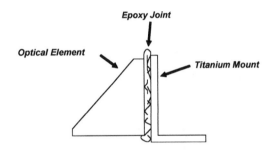

Exhibit 1. Pave VIPER Optical Element and Titanium Mount. The optical elements are bonded to their mounts with an epoxy joint.

The two epoxy agents are mixed when the epoxy is ready to be used. After mixing, the epoxy is carefully applied to the titanium mount and the edges of the optical element that interface with the epoxy mount.

This epoxy mixing portion of the operation is very critical. If too little epoxy is applied, the bond between the mount and the optical element will be too weak, and the titanium mount may release the optical element. If too much epoxy is applied, one of two phenomena will occur: The epoxy will spill over into the optical path (distorting the laser beam and probably burning the optical element), or the epoxy will impart undue stress into the optical element after curing (which will also distort the laser beam).

The Laser Beam Adjustment Process. After the optical elements have bonded to their respective titanium mounts (24 hours are allowed for this process), the titanium mounts are secured in the laser/transmitter. Several of the optical elements' mounts are adjustable to steer the laser beam within tightly-toleranced limits. This adjustment process is performed on an optical bench, which is a tightly-toleranced fixture. A helium-neon laboratory laser (a HeNe laser, referred to as a "Heenee" laser) is used for this adjustment process. A known input from a HeNe laser mounted on the optical bench is directed into the laser, and Omega's laser technicians adjust the laser/transmitter's optical mounts to properly direct the beam (through the laser/transmitter) to a target point, which is also mounted on the optical bench. The optical bench and laser/transmitter are schematically illustrated in Exhibit 2.

Environmental Testing. Once the laser/transmitter assembly operations are complete (the assembly operation includes many additional steps not described above), the completed laser/transmitters are subjected to environmental tests. These tests include thermal shock (the laser/transmitter is placed in an oven for a period of time sufficient to thoroughly heat the device, and then quickly transferred to a freezer), and vibration testing (the laser is tested at specified vibration levels for one hour).

The Failures

Art Bundrock cut power to the vibration table. He was disappointed. As the Pave VIPER reliability engineer, Bundrock was principally responsible for dispositioning Pave VIPER acceptance test failures and for performing the follow-on failure analyses. Several laser/transmitters had recently failed vibration testing. The failures consisted of optical elements tearing away from their titanium mounts during the vibration test.

During most of the Pave VIPER production program, the optical element debonding failures had occurred relatively infrequently. In the last several weeks, however, the number of failures had increased sharply. The first of the recent failures had been discovered during post-vibration laser/transmitter output and boresight testing. In these tests, there had been no laser energy output, and in the post-test teardown, the optical element debonding failures had been discovered.

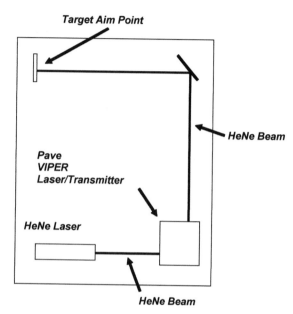

Exhibit 2. *Optical Bench (shown from above) Used for Aligning Pave VIPER Laser Beam Path.*

During recent tests, Bundrock had learned to listen to the laser/transmitter during vibration, and he could now hear the optical elements bouncing around inside the laser/transmitter after they had broken free of their mounts. The tell-tale clattering of glass against metal had prompted him to stop this most recent test.

Having had some experience in systems failure analysis, Bundrock recognized that many potential optical element debonding causes existed. Bundrock also recognized that he needed help to solve the optical element debonding problem. He intuitively felt that something was wrong with either the epoxy or the epoxy application process. Bundrock decided to talk to the laser/transmitter engineers and the laser technicians in the clean room to get their inputs.

The Clean Room Interview

Pave VIPER lasers are manufactured in a clean room environment. Bundrock went to the clean room to speak with Jim Andres, who he felt was the most knowledgeable and skilled laser technician. Andres' duties included aligning the Pave VIPER laser/transmitter optics using the optical bench HeNe laser. Andres had also worked in the optics mounting area (i.e., he had epoxied Pave VIPER optics to their titanium mounts prior to installing the optics in the laser/transmitter). The optics mounting area was also located in the clean room.

Bundrock suited up in a clean room smock, slippers, and a shower cap before entering the clean room. He found Andres adjusting a Pave VIPER laser/transmitter at one of the several clean room benches.

"Good morning, Jim," Bundrock said.

"Good morning to you," Andres answered. "Bet I know why you're here. Still having optics debonding problems?" Andres was familiar with the problem, as the failed laser/transmitters had to be returned to the clean room for a thorough cleaning (the debonded optics tended to break up when they failed in the vibration test, which necessitated a near-complete disassembly of the failed laser/transmitter to clean out the optical element particles, the epoxy chips, and other contaminants created when the optical element bounced around the interior of the laser cavity). Andres was often assigned to clean and then rebuild the failed laser/transmitters.

"Yes," Bundrock said, "and the problem appears to be getting worse. We've had four fail this week. That's more than we deliver in a week."

"Well, I knew you'd be here to see me sooner or later," Andres said.

"Do you have any ideas why these things are coming apart?" Bundrock asked.

"Yep. The epoxy joint is failing during vibration." Andres continued with his work on the laser/transmitter as he spoke. "You're only seeing the three heaviest optics failing...the two corner prisms and the big steering lens. All of the other lighter optics are hanging onto their mounts, or vice versa, I should say. To me that says it's a vibration problem."

"I didn't realize the problem was confined to those optics," Bundrock said. "Why do you think they're falling off?"

"Can't say for sure," Andres answered, "but you might want to look into how the new guys we've got doing the bonding work are doing their jobs. I'm not so sure we're doing the waterbreak test any more. And I caught one of them working on a mount without wearing finger cots, you know, the little rubber finger covers. And even when they do wear the finger cots, I don't know if they've been trained to not touch their face with their fingers. You know, if they do that, the finger cot is contaminated, and it will contaminate the mount and the glass. All a guy has to do is scratch his nose and then touch a mount, and it's going to fail."

"Hmmh..." Bundrock said. He knew that cleanliness was a critical factor in proper epoxy bonding. "Do you think we've got a contamination problem?" he asked Andres.

"I know we do," Andres answered.

The Project Engineer Interview

Bundrock next visited Tom Axelson, the Pave VIPER project engineer. Bundrock explained the problem to Axelson, who was already familiar with it.

"What do you think is causing these failures?" Bundrock asked Axelson.

"We saw similar failures during development," Axelson answered. "They were tied to two causes that I'm aware of. The first was excessive vibration, and the other was contamination."

"What do you mean about excessive vibration?" Bundrock asked.

"I mean that the vibration platform was vibrating in excess of the spectrum specified in the Acceptance Test Procedure. The g-levels were about 30 percent higher than they should have been, and the heavier optics were falling off left and right."

"What was done to fix that problem?" Bundrock asked.

"Not sure I remember now," Axelson answered. "It's been about three years or so. I think we started calibrating the vibration table, but I'm not sure. Calibration could probably tell you about that. You need to talk to Bill Olson in the Test Department. He'd remember more about the development failures."

The Test Engineer Interview

Bundrock's next stop was to see Bill Olson. Olson had also heard about the optics debonding failures, as they occurred in his area of responsibility.

"Bill, Tom Axelson sent me over to see you," Bundrock said. "I know you know about the development failures where the vibration tables were out of spec. Do you think we still have that problem now?"

"No way," Olson answered. "We keep very tight control over our test equipment. We've never had an optics debonding failure since development, which was before my watch anyway. I know our equipment is not causing the optics debonding failures."

"I'm sure that's right," Bundrock said. "I just wanted to learn a little more about how your equipment is calibrated so I can rule this out as a potential cause of failure. How often do you calibrate the vibration table?"

"You'd have to check with Calibration, over in Quality Assurance, for that information," Olson said.

"Okay," Bundrock said. "Could you show me how you control the vibration table, just so I could understand it a little better?"

"I can't do that right now," Olson answered. "After the last failure this morning, my technician and I knew we'd have some down time, so we're doing a little preventive maintenance on the panels

right now. They're disassembled, so there's really nothing to see, anyway."

"Do you operate the vibration tester?" Bundrock asked.

"No, the technicians do that," Olson said.

"Is it the same one every time?" Bundrock asked.

"Of course not," Olson answered. "We use whoever's available at the time. All of my technicians are checked out, though. They know how to operate the equipment."

"I see," Bundrock answered. He thought for one last moment before leaving the test area. "Were there any recent changes to the test equipment?"

"None that would influence these failures," Olson said. "We changed the mounting block for the laser/transmitters about two months ago, but that wouldn't have any affect on the vibration table output."

"Hmmh..." Bundrock said. "When did you say the was the last time you calibrated the test setup, Bill?"

"Well, I'm not sure," Olson said. "I think it was about six months ago. The table only has to be calibrated once a year."

"Okay, Bill, thanks for helping me out," Bundrock said. "Have you got any ideas on what could be causing these debonding problems?"

"Not really," Olson answered. "I only know that whatever it is, it's not in my area."

The Quality Engineer Interview

Bundrock's next stop was in Brian Enod's office. Enod was the Pave VIPER program quality engineer.

"Brian, I've talked to Bill Olson in Test, Jim Andres in the clean room, and Tom Axelson over in Engineering," Bundrock said. "I'm trying to get to the bottom of these optics debonding failures."

"I've been looking into those, too," Enod said. "I can guess what you got from the folks you've talked to already. Test said it wasn't a test problem, engineering said it wasn't a design problem, and manufacturing said it wasn't a process problem, right? You'd almost have to guess that the failures haven't been occurring."

"No, not really," Bundrock said. "Most of the guys were pretty open about what goes on in their area. In fact, most of the questions I have for you are based on inputs from them."

"I'll do whatever I can to help," Enod said.

"First question, then," said Bundrock, "is what is the waterbreak test? Andres, in the clean room, mentioned that to me."

"That's a test where we spray a very fine water mist onto a clean optic to determine how clean it really is," Enod began to explain. "The theory behind the test is that if the item we're inspecting is clean, the water will run right off. If it's dirty, the water will bead up. We use a special device called a goniometer to measure the size of the water beads on the moistened optic, and if they fall below a certain level, the optic meets its cleanliness requirements."

"I think I understand that," Bundrock said. "Are we actually doing that test now?"

"There was some talk about taking it out of the planning," Enod said. "I'm not sure where we are on that issue right now."

"What about calibration on the vibration tester, Brian?" Bundrock asked.

"Well, that's another sore point," Enod said. "You know how Olson is. Nothing ever goes wrong in his area. The equipment is supposed to be calibrated, but I've never understood how we calibrate a vibration test fixture, and Olson is always taking that thing apart and putting it back together. Fact is, he scares Merrit, the calibration guy, so Merrit doesn't go over there unless he absolutely has to, which is about once a year. You're wondering, I guess, if the fixture is

overvibrating the lasers, and the answer to that is probably not."

"How do you know that?" Bundrock asked.

Enod thought silently for a moment. "I guess I don't," he said.

"What else could be giving us grief, here, Brian?" Bundrock asked.

"Well, we've always had problems with the epoxy," Enod said. "Sometimes the guys don't mix it in exactly a one-to-one mix. Sometimes they don't stir it enough and we get localized improper mixing. Sometimes they stir it too hard and make bubbles, which, of course, weakens the bond. Sometimes they don't use it within the thirty-minute window they're supposed to before it begins to set. Once the epoxy begins to set, you shouldn't apply it to any surfaces to be bonded. You're supposed to apply it to the optic and the mount before it starts to set. And that's another problem. Sometimes the guys don't apply it to both surfaces, you know, the mount and the optic. They just slop it on one and figure it will contact the other. I think we could have problems because we may not wait the full 24 hours for the epoxy to cure, too. Sometimes the laser technicians get in a hurry, and they start the optical bench alignment process, you know, with the HeNe laser, before the optics are dry. If they're disturbed before the epoxy has fully cured, that will also weaken the bond."

"What else?" Bundrock asked.

"The big problem has always been contamination, at least in my opinion," Enod said. "You know, you can tell by looking at a failed optic if it was dirty or if the epoxy had already begun to set up when it was applied."

"How so?" Bundrock asked.

"If the epoxy makes a clean break from either the mount or the optic, with no epoxy left on one of the mounting surfaces, then the surface wasn't clean, or the epoxy had already started setting up before it was applied to the surface," Enod explained. "If there is epoxy on both surfaces, it's likely it failed in the epoxy joint itself, and not at the epoxy interface. If that happens, the epoxy failed structurally, and that means that either the epoxy is too weak for the loads it sees, or the loads were too great."

"Like if the laser was overvibrated?" Bundrock asked.

"Potentially," Enod said. "You know, there's one other thing that comes to mind, and that's that the epoxy was too thin."

"Do you mean the mix ratio was wrong?" Bundrock asked.

"No," Enod said, "I mean the joint was just too thin. The laser techs are supposed to maintain a minimum epoxy thickness between the mount and the optic, and if you go below that, the joint is structurally weakened."

"How do we control that?" Bundrock asked. "Do we use a special tool or something?"

"Good question," Enod said.

Questions

1. To learn more about the optics debonding failures, Bundrock interviewed several of the people involved in producing and evaluating the Pave VIPER laser/transmitter. What are your comments on how Bundrock went about gathering this information?

2. What else might Bundrock have done to learn more about the optical element bonding process?

3. What else might Bundrock have done to learn more about the optics debonding failure history?

4. Who else should Bundrock meet with to learn more about the optics debonding failures?

5. Prepare a fault tree analysis for Pave VIPER optics debonding.

6. Prepare a failure mode assessment and assignment matrix for the above fault tree.

Index

J.H. Berk and Associates Seminars - Available Training Programs

Training programs offered for in-house presentation by **J.H. Berk and Associates** include systems failure analysis, continuous improvement technologies, risk management, quality measurement systems, value improvement, technical writing, design of experiments, manufacturing leadership training, and engineering drawing interpretation. Other tailored programs can be developed on short notice to meet client needs. Detailed descriptions of our training programs are included in the following paragraphs. Please contact us at www.ManufacturingTraining.com for additional information.

Managing Effectively. The *Managing Effectively* seminar focuses on the needs of new managers or others requiring basic management training. Topics include planning, budgeting, cost tracking, interviewing, delegating, promoting, counseling, terminations, and other global tasks (problem solving, managing conflict, and evaluating and controlling risk). The program also includes an overview of popular software packages used for planning, budgeting, and other management tasks. *Managing Effectively* is offered as a three-day workshop or a six-class course.

Systems Failure Analysis. The *Systems Failure Analysis* seminar integrates Engineering, Quality Assurance, Manufacturing, Procurement, and other organizations' efforts to identify and eliminate the root causes of failures occurring in complex systems. The approach utilizes fault tree analysis for identifying all potential failure causes. Hardware analysis, statistical analysis, design of experiments, technical data package evaluation, and other technologies are combined to define root causes and select optimal corrective actions. Industry-based case studies are utilized extensively, and during the course students analyze failures occurring in their products. *Systems Failure Analysis* is offered in two- or four-day workshops, or in a ten-week (three hours per class) format.

Risk Management. The *Risk Management* seminar focuses on basic program, product, and process risk management techniques. It explores issues associated with consumer and producer risk, warranty evaluation and costing, failure modes and effects analysis, yes-no-risk assessments, system safety analyses, industrial safety, and other risk management technologies. *Risk Management* is offered in half-day and one-day workshops.

Quality Measurement Systems. The *Quality Measurement Systems* seminar explores continuous improvement technologies related to nonconformance documentation, data analysis, and corrective action implementation. Nonconforming material report formats, data base selection, Pareto charts, quality improvement newsletters, cost of quality, and other statistical quality control reporting systems are described. *Quality Measurement Systems* is offered in one-day and two-day workshops.

Value Improvement. The *Value Improvement* program of instruction develops a management system for systematically improving value by identifying value improvement opportunities. The program identifies value improvement opportunities in the areas of material usage, scrap and rework reduction, packaging improvements, reusable assembly aids, process improvements, enhanced procurement philosophies, product design simplifications, transportation improvements, inspection and test redesign, improved warranty designs, and material and paperwork movement reductions. *Value Improvement* is offered in a three-day workshop.

Technical Writing. The *Technical Writing* seminar focuses on effective technical writing for engineers, managers, manufacturing and quality assurance personnel, and other business professionals. The *Technical Writing* program includes a review of basic grammar requirements as well as more advanced business communications concepts, including word processing fundamentals, generating and importing electronic images and other data, and other topics related to style and contemporary written communications practices. The *Technical Writing* program also addresses the special needs of technical and managerial personnel who have English as a second language. The *Technical Writing* course is offered in half-day, one-day, and three-day formats.

Design Of Experiments. The *Design of Experiments* seminar focuses on applying statistical concepts and analysis of variance techniques to product and process design. The nature of experimentation, correlation and regression, Taguchi techniques, surface response methodologies, and sampling techniques are addressed. The design of experiments course is offered as a three-day workshop, or in an eight-class (three hours per class) format.

Manufacturing Leadership Training. The *Manufacturing Leadership Training* is a 15-class, 30-hour program of instruction focused on developing manufacturing lead person leadership skills. The *Manufacturing Leadership Training* program provides manufacturing and quality assurance lead personnel with the ability to understand lead person responsibilities and authorities, develop and use leadership skills to motivate their teams, drive the organization in implementing and adhering to MRP II or Kanban requirements, communicate effectively with subordinates, peers, and superiors, identify and resolve conflicts, and understand and use productivity measurement techniques. The program also emphasizes recognizing and acting on employee training needs, and implementing a Total Quality Management philosophy throughout the organization. The training approach includes lectures, interactive discussion, training exercises, and a final examination.

Engineering Drawing Interpretation. The *Engineering Drawing Interpretation* course is a 10-class, 20-hour program of instruction focused on developing manufacturing and quality assurance technician abilities to read and understand engineering drawings. The *Drawing Interpretation* program provides manufacturing and quality assurance personnel with the ability to understand basic engineering drawing concepts and definitions (including drawing title blocks, views, datums, zones, notes, parts lists, and other features), interpret basic dimensions and tolerances (including geometric dimensioning and tolerancing), and understand basic configuration management concepts. The *Engineering Drawing Interpretation* program can be presented on site either after hours or prior to the start of the work day in two-hour sessions, one day per week, for 10 weeks. The training approach includes an initial drawing interpretation skills assessment, lectures, interactive discussion, training exercises, and a final examination.

J.H. BERK AND ASSOCIATES
1082 PEPPERTREE LANE
UPLAND, CALIFORNIA 91784
(909) 265-5932 (TELEPHONE)
(909) 982-3022 (FAX)
WWW.JHBERKANDASSOCIATES.COM